Landforms
an introduction to geomorphology

Ian Galbraith and Patrick Wiegand

Oxford University Press

Contents

Oxford University Press, Walton Street, Oxford OX2 6DP

Oxford   London
New York   Toronto   Melbourne   Auckland
Kuala Lumpur   Singapore   Hong Kong   Tokyo
Delhi   Bombay   Calcutta   Madras   Karachi
Nairobi   Dar es Salaam   Cape Town

and associated companies in
Beirut   Berlin   Ibadan   Mexico City   Nicosia

Oxford is a trade mark of Oxford University Press

First published 1982
Reprinted 1983 (twice), 1984
ISBN 0 19 913271 2

Phototypeset by Tradespools, Frome, Somerset

Printed in Hong Kong

Introduction

We hope that you will find this book useful if you are preparing for an examination and also that it will encourage you to look more closely at the landforms around you. Wherever you live or travel, the earth's surface provides varied scenery which becomes more interesting if you understand how it was formed. Ideally you should study landforms at first hand but unfortunately opportunities for field work are always limited. Photographs and diagrams are the next best thing and we believe that careful study of the illustrations is just as important as reading the words in between!

1 Landforms and landscapes

Fig 1 Snowdonia: a landscape made up of landforms

Look at the photograph (Fig. 1) of Snowdonia. This impressive *landscape* is made up of *landforms* such as mountain peaks, ridges and lakes. Other landforms, not shown here, include volcanoes, waterfalls, cliffs and sand dunes. Landscapes are made up from different combinations of landforms.

The science that studies landforms is called *geomorphology*. Geomorphologists are interested in the shape of landforms, the processes that make them the shape they are and how their shape has changed through time.

1 Describe the landscape of your home area. If you live in a town try to describe the shape of the land on which the town is built. You should be able to identify hills and valleys even though they may be heavily disguised by buildings. You may find it helpful to refer to the contours on a local Ordnance Survey map.

It can be difficult to describe the shape of the land. One way of describing landforms is to make a map. The contours on an Ordnance Survey map give a general impression of the differences in height or *relief* of the surface. However, this is not usually detailed enough for geomorphologists. For them a *morphological map* is a more scientific way of mapping landforms. This kind of map uses symbols to show important details of landforms such as changes (or 'breaks') in the slope of the land which may be too small to be shown by contours.

2 Fig. 2 shows Glenarrif in County Antrim, N. Ireland. Make a copy of Fig. 3 which is a base map of the area shown in the photograph. Only the area within the arrows can be seen. Part of the morphological map has been started. Complete the map using the symbols shown.

Fig 3 Morphological map of Glenarrif

Fig 2 Glenarrif, County Antrim, N. Ireland

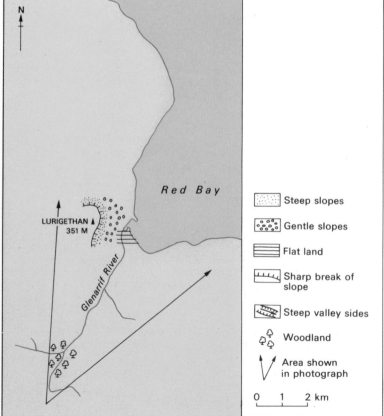

2

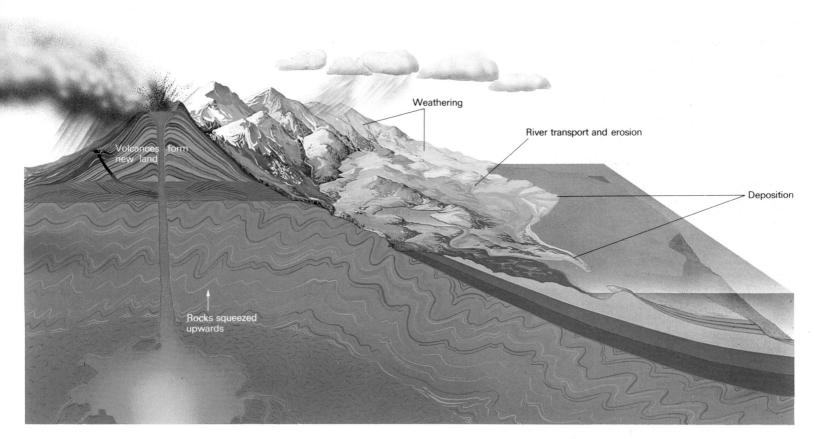

Fig 4 Mountain building and denudation

The following labels appear on the figure:

- Weathering
- River transport and erosion
- Deposition
- Volcanoes form new land
- Rocks squeezed upwards

Landform processes

A *process* is a series of events or changes. Landscapes are the result of two major sets of processes acting on the surface of the earth (See Fig. 4).

Firstly, there are those processes which add new material to the earth's crust or which cause it to be uplifted. These include the flow of molten lava from volcanoes and large scale earth movements. We may call these the processes of *mountain building*.

Secondly, there are those processes which destroy or wear away the rocks and landforms of the earth's surface. These are called the processes of *denudation*.

Rocks on the earth's surface are in contact with the atmosphere and the oceans and it is water that is primarily responsible for denudation. Rain water causes rocks to decay, valleys are cut by running water and by glacier ice and the coast is worn away by the action of waves. It is easier to understand denudation if we see it as the result of several distinct operations.

Weathering is the breaking down of rocks at or near the surface of the earth. This word is used because it is the weather that is mainly responsible. Some minerals in rocks for example are dissolved by rainwater which causes the rock to crumble.

Rock waste, broken up by weathering, does not remain in the same position very long. Rivers, glaciers, the sea and winds carry weathered fragments away. This is called *transportation*. Transportation may also be a direct result of gravity. This occurs when weathered fragments roll downhill or fall from a cliff face. Material moved by all these processes may be large boulders of solid rock or tiny particles dissolved in water.

The word *erosion* refers to the cutting and shaping of the earth's surface. Think of a sculptor carving a statue. He uses a chisel to break off pieces of wood or stone. These pieces fall to the floor by gravity. In the landscape running water, the sea, moving ice and wind act as chisels. They transport rock fragments and also use these fragments to wear away the landscape. Pebbles at the bottom of a cliff for example are transported by the waves but they may also be hurled at the cliff in stormy seas, eroding it.

All transported rock particles eventually settle. This is called *deposition*. Rivers deposit fine silt and mud in the sea, sometimes as deltas, Similarly, glacier ice dumps the rocks it carries when the ice melts. The rocks broken off crags pile up in heaps at the foot of the mountainside.

If conditions stay the same for long enough these deposits may become compressed by the weight of more deposits on top. Eventually this results in new rock being formed.

Fig 5 Making a model river basin with a stream table

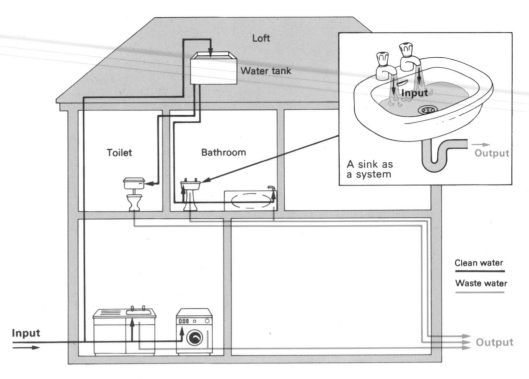

Fig 6 A water system in the home

One way of learning more about the processes at work on landscapes is by making models. The picture above shows a *stream table*. This is a large tray filled with sand. Water squirted through a jet flows across the sand and behaves like a river. This smaller version of a river allows us to simplify the real river so that we can understand the basic processes at work.

Another way of helping us to understand processes in geomorphology is by thinking of them as *systems*. Think of the appliances that use water in your home. The sink, bath, w.c. and so on are all separate but you cannot really understand how they work unless you think of them all being linked together. Fig. 6 shows the water system of a house. Only the cold water is shown to keep the diagram simple. Notice that the system depends on a flow through the pipes in the house. Pure water going into the house is called the *input* and the waste water is called the *output*.

Systems can be of different sizes. One of the smallest water systems in a house might be a sink. Input is through the taps and output through the waste pipe. The water system diagrams can help us to understand how the water supply in a home *works*. We can understand how the landscape *works* by drawing similar system diagrams.

Fig. 7 shows the largest water system of all – the *hydrological (or water) cycle*.

3 Make a copy of Fig. 7.
Describe what happens to water as it passes from one part of the cycle to another. Start with water evaporating from the oceans.

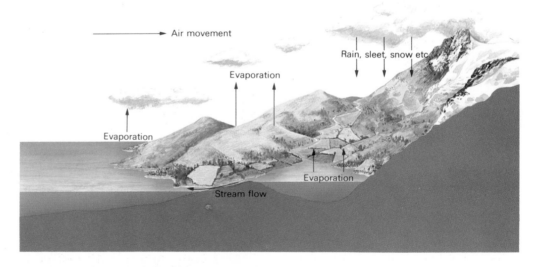

Fig 7 The hydrological cycle: the biggest water system

This particular system is very large, It represents the movement of all the water on the earth. There are no inputs or outputs as all the water in the world is recirculated within the cycle. Systems in the landscape are not all large however. We could look at a river basin as a smaller part of the larger system (like the sink in the house). A river basin would have an input of water from rainfall and an output of water along the river and by evaporation.

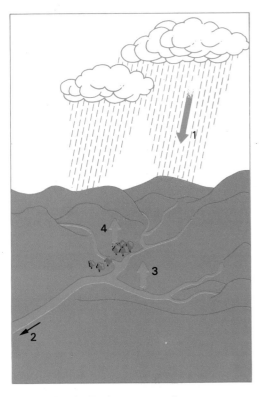

Fig 8 A river basin shown as a system

had a much longer history – perhaps 500 million years compared to one thousand for the cathedral. Time is therefore an important aspect of the study of geomorphology. Many of the processes operating on the surface of the earth are very slow. It is not surprising therefore that men used to think of the landscape as unchanging – the hills and valleys were seen as 'everlasting'.

We now know that a great many changes have taken place in the past to explain the present landscape and that these changes, however gradual, are continuing today. The position of land and sea has altered many times and the climate has also changed. The present landscape of Britain is only temporary.

Geomorphology and people

Why should we study geomorphology? One reason is that many people are naturally interested in their surroundings. Every year, more people are spending their leisure time in the countryside and want to understand the scenery they enjoy. The physical environment is also important as a background to the study of human geography as the landscape partly explains the use man makes of the earth.

Geomorphology is also a useful science. The more that is known about landforms and the processes that cause them to change, the more we are able to apply that knowledge for our benefit. Geomorphologists may be involved in research into the causes of landslides or flooding. Or they may be asked to advise on the problems of subsidence or the protection of coastal areas.

5 People are also important as a force modifying the landscape. There are many 'man-made' landforms. Be careful not to confuse them with natural ones. Look at the photograph of Silbury Hill. It is not a natural feature but a Bronze Age burial mound, the largest artificial hill in Europe.
What other 'man-made' landforms can you think of? Are they features of 'erosion' or 'deposition'?
Can you think of examples where modern technology has enabled people to modify the landscape even more than at Silbury Hill?

4 Make a copy of the diagram of a river basin. Label the arrows showing the movement of water and say whether each is an input or output.

Systems like this help us to simplify the real world in the same way as the stream table does. For this reason we can also call systems *models*.

Time

Consider a large cathedral such as Canterbury or St. Paul's. During its long history it may have been much altered from its original style and shape. Some parts have worn down slowly through age or have been destroyed suddenly in war time. Builders throughout history have added new parts to the original building perhaps in different styles or using different materials. The building we see today is therefore the product of a series of changes in its history. The landscape is like this only more complex. This is partly because it has

Fig 9 Silbury Hill: a 'man-made' landform

2 Rocks

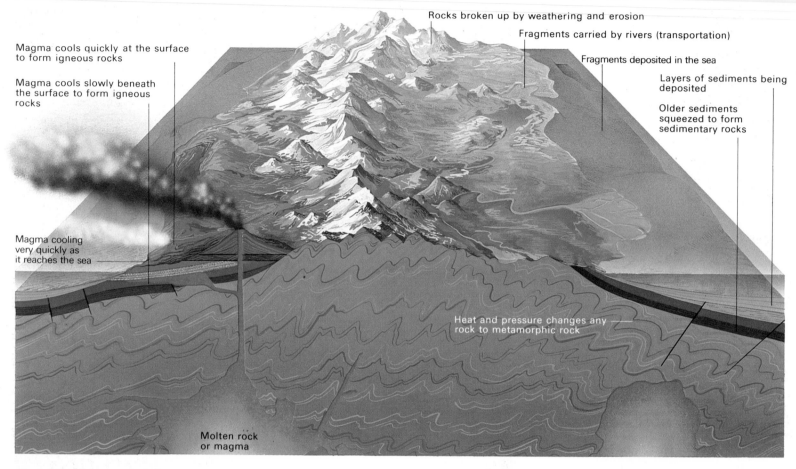

Magma cools quickly at the surface to form igneous rocks

Magma cools slowly beneath the surface to form igneous rocks

Magma cooling very quickly as it reaches the sea

Molten rock or magma

Rocks broken up by weathering and erosion

Fragments carried by rivers (transportation)

Fragments deposited in the sea

Layers of sediments being deposited

Older sediments squeezed to form sedimentary rocks

Heat and pressure changes any rock to metamorphic rock

Fig 1 Another system: the geological cycle

Many different rocks make up the crust of the earth. They may be hard like granite, soft like clay or loose like gravel. There is a great variety of colour, weight and hardness. It is worth knowing more about the study of rocks in order to understand landforms.

Rocks are made up of *minerals*. If you think of a rock as being like a fruit cake then the minerals would be represented by the currants, sultanas and cherries together with the egg, flour and milk that bind them all together.

Figure 3 shows you a mineral that has grown quite large. *Quartz* is a glassy mineral that is often milky white. When pure it is known as rock crystal. Purple quartz is known as amethyst. Quartz is hard – it cannot be scratched with a knife.

Some minerals contain metals and are called *ore minerals*. The word *ore* is usually only used for a mineral that is worth mining. Very hard minerals are often cut, polished and used for jewellery, particularly if they have an attractive colour and are free from flaws. If they are rare they are called *precious stones* or *gems*.

1 Make a drawing from Fig. 3 of the crystals of quartz. How many sides do most of the crystals have?
2 Study the list of minerals below. Some are ores, some are precious stones. Use a dictionary or encyclopaedia to help divide them into the two groups. For each ore state the metal obtained and for each precious stone give its colour.

diamond	haematite
chalcopyrite	ruby
galena	emerald
opal	bauxite

Rocks are usually grouped into three types, depending on how they were formed. These three groups are:

a the *igneous* rocks
b the *sedimentary* rocks
c the *metamorphic* rocks

The relationship between these three rock types may be seen above. This diagram shows a system called the *geological cycle*. Remember that it would take millions of years for the materials that make up the crust of the earth to pass through this cycle.

Fig 2 A geologist studying igneous rocks: a) granite   b) basalt under magnification   Magnification ✕ 16·5

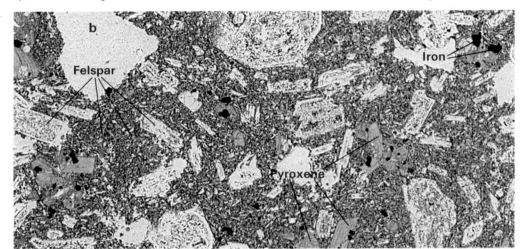

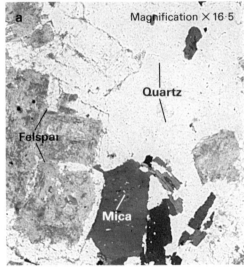
a   Magnification ✕ 16·5

Felspar

Quartz

Mica

Fig 3 Quartz

Igneous rocks

These are rocks that have cooled from the hot liquid or *magma* beneath the crust of the earth. If movements take place in the crust this magma may push upwards and spill out on the surface as volcanoes or lava flows. As the magma cools *crystals* grow and igneous rock is formed. Large crystals are formed if the magma cools slowly. Magma on the surface cools rapidly though and the crystals are small. Igneous rocks can usually be recognised by their crystals and the minerals may be especially clearly seen if the magma cooled slowly.

3 Study Fig. 1. Magma is cooling in three different ways. Make a copy of the diagram and add brief notes to describe the size of the crystals that will be found in the rocks when the magma has cooled.

4 A geologist has collected two types of igneous rock. To help her examine them she has made a thin slice of each rock and studied these carefully under a microscope. She can then see the minerals clearly.
Study Fig. 2 which shows the rocks she has collected and the thin slices. If possible study actual specimens of the rocks themselves. For each rock:

a say whether it is light or dark
b say what minerals form it.
c describe the colour, shape and size of the minerals.

The geologist concludes that rock A is *granite* and that rock B is *basalt*. Notice that the size of many of the crystals in basalt is very small. Under what conditions do you think granite and basalt cooled from the molten state?

When igneous rocks cool they often shrink and crack. These cracks are known as *joints*. Fig. 32 in Chapter 3 shows the hexagonal pattern of joints that is sometimes found in basalt.

7

Sedimentary rocks

Rocks at the surface are broken up by the weather. Small particles of rock are often carried away by water, moving ice or wind and deposited somewhere else. The deposited material is called a *sediment*. The most common types of sediment are found under water in the sea. The layers of soft, loose mud and sand that have been deposited by rivers are gradually changed into sedimentary rocks. After many years successive layers of sediment accumulate on the sea floor. Air and water are squeezed out of the lower layers. Often the water that is squeezed out leaves behind chemicals that may cement the grains of mud and sand together. *Sandstone* is a sedimentary rock. Most of the rock is made up of 'sand' or grains of quartz.

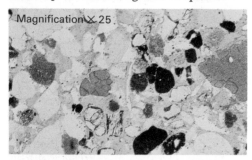

Magnification × 25

0 2 4 cm

Fig 4 Sandstone: a sedimentary rock. Under magnification (above)

5 Study the figure above and, if possible, a piece of sandstone. Notice that the grains of quartz are more rounded in the sandstone than in granite.

a Can you explain why? Remember where the quartz has come from in each case.

b The hardness of quartz can be compared with the other two minerals in granite (felspar and mica) as follows:

 quartz: cannot be scratched with a penknife

 felspar: can be scratched with a penknife

 mica: can just be scratched with a finger nail

Consider, from the beginning, how sedimentary rocks are formed. Can you suggest why many of them are made of quartz and not of felspar or mica?

6 Half fill a glass container with water. Drop a handful of sand into the container and allow to settle. Drop in another handful of powdered clay and leave to settle for a few days. Repeat the procedure. You will find that *layers* are settling in the container.

These layers represent different periods of deposition. The layers of rock or *strata* in sedimentary rocks may often be clearly seen, separated by *bedding planes*.

When rocks are deposited the strata and bedding planes are usually horizontal. Later though, because of movements in the earth's crust, the rocks may be tilted and the strata are then said to *dip*. The angle of dip may be measured easily using a *clinometer* – an instrument rather like a protractor.

Sedimentary rocks are often divided into groups according to the size of the fragments that make them up. *Conglomerates* are made from rounded pebbles. *Clay* is a sedimentary rock with very small grains – about 1/250 mm in diameter. *Sandstone* has grains larger than clay but smaller than conglomerate.

Not all sediments, though, are formed from the remains of other rocks. There is a second type of sedimentary rock which consists of the remains of plants or animals or the accumulation of chemicals in water. Where the sea contains more of a chemical than it can dissolve the surplus is deposited on the sea floor. Calcium carbonate, rock salt and gypsum are sometimes deposited directly in this way. Calcium carbonate can also be formed by the action of sea creatures in making their shells. When the animals die their shells fall to the sea bed

Fig 5 Shelly limestone

and pile up, often to a great thickness. Masses of shells may then be cemented together, usually by more calcium carbonate. The shells are often clearly seen in the rock (see above). A rock which contains a large amount of calcium carbonate, whether it has been deposited directly or indirectly in the form of shells is known as a *limestone*.

Fossils are the traces of animals and plants that have been preserved in rocks. There are several types of fossil. Firstly the shell or skeleton of the animal may be preserved – as in Fig. 6. Secondly, the shell or skeleton may have left an impression in the mud in which it was deposited. Thirdly, the original plant or animal may have been gradually replaced by some other substance. Often, for example, crystals of a mineral called *calcite* grow in the spaces left by shells that have rotted. Fig. 6 shows how a fossil is formed.

Dead ammonite on the sea floor. These animals looked rather like squids and are now extinct.

The fleshy parts have now rotted, and the spiral shell is being covered in mud.

The shell is now buried. It may decay and be replaced by a mineral such as calcite.

Fig 6 The formation of a fossil ammonite

5 CM

Fig 7 A fossil echinoid

7 Look at the fossil above. It is an *echinoid* or sea urchin that lived about 150 million years ago. The spines that once covered it are no longer to be seen and only the cast of the hard skeleton now remains.
Write out the sequence of events leading to the preservation of the fossil, starting with it being alive.

8 Most fossils are the remains of animals with shells that lived in the sea. Why should this be so?

9 Fossils are only found in sedimentary rocks. Why are there no fossils in igneous rocks?

Not all sedimentary rocks are formed under the sea. New Red Sandstone was formed at a time when much of Britain was a desert by the cementing together of wind blown sand. If you look at New Red Sandstone under a magnifying glass you can see that the small grains of quartz are scratched. This scratching happened as the grains jostled against each other in desert sandstorms.

Metamorphic rocks

Metamorphic rocks are those that have been changed from their original form by heat or by pressure beneath the surface of the earth.

When molten rock, at a temperature of several hundred degrees centigrade, is forced upwards the surrounding rock becomes baked and hot gases are given off. New minerals and crystals grow and the rock is altered.

Slate is a metamorphic rock which has been formed by the prolonged heating and squeezing of *shale*, a soft sedimentary rock formed by the deposition of mud. Slate splits easily into thin sheets and for this reason has been important as a roofing material. Fig. 8 explains why the rock does this.

10 Under heat and pressure limestone becomes *marble* and sandstone becomes *quartzite*.
a What is marble commonly used for?
b Metamorphic rocks are often very hard and resistant to wear. Why should this be so?
c Why are few fossils or bedding planes present in sedimentary rocks that have experienced heat or pressure?

11 Refer to the diagram of the geological cycle (Fig. 1). Make a copy of the diagram and write a paragraph to describe how the cycle works.

Shale

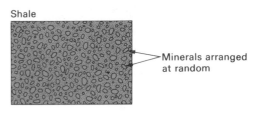

Minerals arranged at random

Shale is squeezed

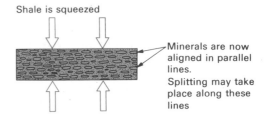

Minerals are now aligned in parallel lines.

Splitting may take place along these lines

Fig 8 How slate is formed

Geological time

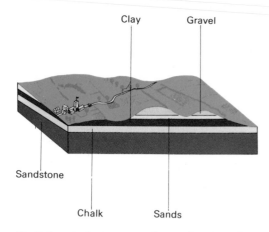

Clay   Gravel

Sandstone

Chalk   Sands

Fig 9 A geological cross-section: a slice through the earth

12 Look at the geological cross-section above.

Are the rocks in the diagram igneous, sedimentary or metamorphic?

Write down the order in which they were deposited, starting with the oldest.

You probably assumed that the oldest rocks were the lowest and that the most recent rocks were on top. The relative age of rocks is in fact usually worked out in this way.

Fossils are also used to decide the age of rocks. Plants and animals have lived on the earth for about 600 million years. During this time they have gradually changed their form to suit different environments. This process is called *evolution*. Strata of the same age contain fossils at the same stage of evolution wherever they are found. We know therefore that strata from different regions are the same age if they contain the same fossils.

The earth is about 4600 million years old but we know very little about the earliest rocks. Most rocks in Britain are less than 500 million years old. Geological time is so long that it is helpful to divide it up into periods – rather like periods in history. The names of the periods are shown in the Geological Column on the right. Periods are grouped together into longer spans of time called *eras*.

Fig 10 The geological column

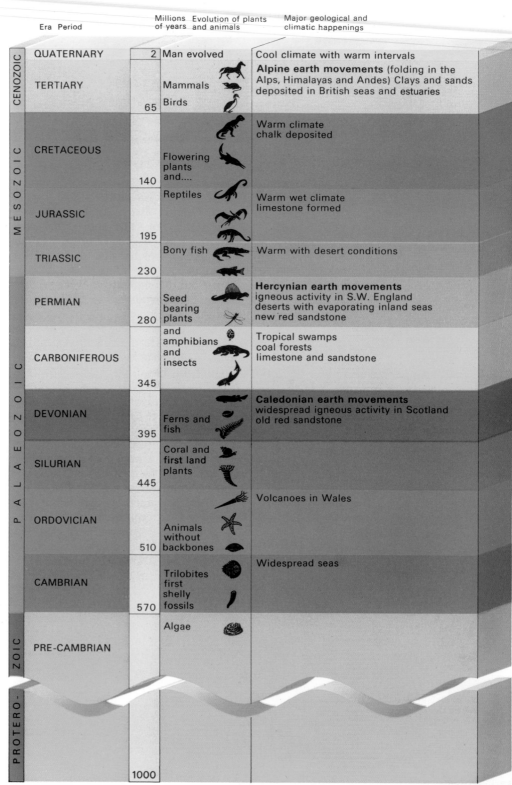

Era	Period	Millions of years	Evolution of plants and animals	Major geological and climatic happenings
CENOZOIC	QUATERNARY	2	Man evolved	Cool climate with warm intervals
CENOZOIC	TERTIARY	65	Mammals, Birds	**Alpine earth movements** (folding in the Alps, Himalayas and Andes) Clays and sands deposited in British seas and estuaries
MESOZOIC	CRETACEOUS	140	Flowering plants and....	Warm climate chalk deposited
MESOZOIC	JURASSIC	195	Reptiles	Warm wet climate limestone formed
MESOZOIC	TRIASSIC	230	Bony fish	Warm with desert conditions
PALAEOZOIC	PERMIAN	280	Seed bearing plants	**Hercynian earth movements** igneous activity in S.W. England deserts with evaporating inland seas new red sandstone
PALAEOZOIC	CARBONIFEROUS	345	and amphibians and insects	Tropical swamps coal forests limestone and sandstone
PALAEOZOIC	DEVONIAN	395	Ferns and fish	**Caledonian earth movements** widespread igneous activity in Scotland old red sandstone
PALAEOZOIC	SILURIAN	445	Coral and first land plants	
PALAEOZOIC	ORDOVICIAN	510	Animals without backbones	Volcanoes in Wales
PALAEOZOIC	CAMBRIAN	570	Trilobites first shelly fossils	Widespread seas
PROTEROZOIC	PRE-CAMBRIAN	1000	Algae	

It is difficult to imagine the length of time that has passed since the earth was formed. However, we could compare the history of the earth with a 24-hour day. If the earth was formed at midnight last night, the first simple life would not appear until 9.00 p.m. this evening. Mammals would not evolve until a quarter to midnight tonight and the whole of our history would take place in the last half minute of the day.

13 Answer the following questions using the geological column, the simplified geological map of Britain on the right and an atlas.

a The line A–B on the geological map divides Britain into two. Which two estuaries does the line join?

b Put the following phrases into two columns, one headed 'North and West', the other headed 'South and East'.

igneous, metamorphic and hard sedimentary rocks
softer sedimentary rocks
Highland Britain
Lowland Britain
rocks mainly of Carboniferous age and older
rocks mainly younger than Carboniferous age

c What is the difference in geology
i) between Dartmoor and Exmoor?
ii) between the Cotswolds and the Chilterns?

d Of which rock or rocks are the following areas formed?
the Pennines
the Antrim Plateau
the Grampians
the Fens

e Give the names and approximate heights of four ranges of chalk hills.

f Write the names and the approximate date before the present of the three major periods of earth movements that have affected Britain.

g In which geological periods were the New Red Sandstone and the Old Red Sandstone deposited?

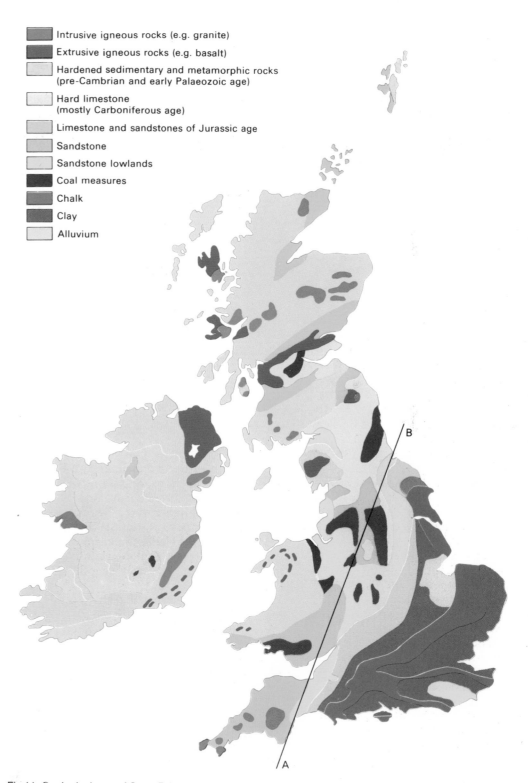

Intrusive igneous rocks (e.g. granite)
Extrusive igneous rocks (e.g. basalt)
Hardened sedimentary and metamorphic rocks (pre-Cambrian and early Palaeozoic age)
Hard limestone (mostly Carboniferous age)
Limestone and sandstones of Jurassic age
Sandstone
Sandstone lowlands
Coal measures
Chalk
Clay
Alluvium

Fig 11 Geological map of Great Britain

The Carboniferous period

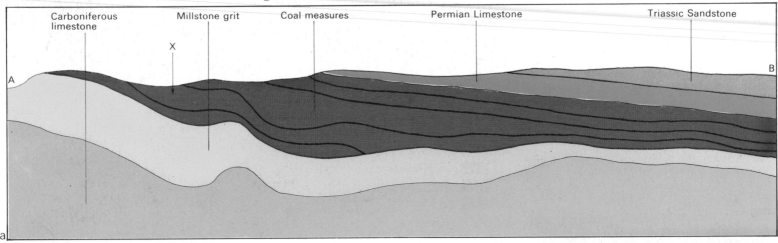

Fig 12 The Yorkshire, Derbyshire and Nottinghamshire Coalfield a) Cross-section

b) Map

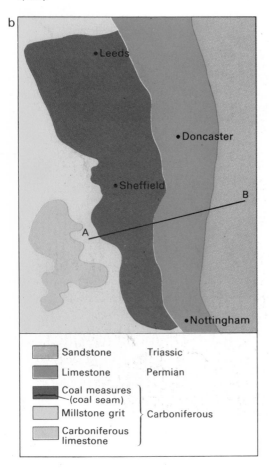

Sandstone — Triassic

Limestone — Permian

Coal measures
(coal seam)

Millstone grit — Carboniferous

Carboniferous
limestone

As it is not possible to study all the geological periods in detail, one is selected for special examination. The word 'carboniferous' means 'containing coal' and although other rocks were deposited in this period it is coal that makes the period especially important to Britain. This can be illustrated by comparing the geological map with an atlas map of population of the British Isles. What similarities and differences are there? Can you explain them?

The shape of Britain today has only been formed in comparatively recent geological time. Our present land area has been at various times covered by sea, mountains, desert, tropical forest and volcanoes. During the Carboniferous period much of Britain was a shallow sea into which sediments were gradually deposited.

14 Study the map and cross section of part of Yorkshire and the East Midlands (Fig. 12). Find this area on your atlas map of Britain. Carboniferous rocks were deposited here with younger rocks on top. Note that the rocks have been bent by earth movements since they were laid down.

a Write the names of the Carboniferous rocks, starting with the oldest.

b What is the approximate age of the Carboniferous rocks?

c Study the photograph of the fossil that was found in the coal measures at X.

Fig 13 A fossil from the Carboniferous period

Describe its appearance or draw it. What kind of plant does it look like?

d What evidence suggests that the Carboniferous rocks are sedimentary rocks?

In the early part of the Carboniferous period the part of the earth's surface we call Britain was much nearer the equator. It has gradually moved to its present latitude over the last 350 million years or so. (The way in which parts of the earth's surface may move is described in the next chapter.) At the beginning of the Carboniferous period only part of 'Britain' was land – the north of Scotland and an island near to our present 'Midlands'. This island has been given the name 'St. George's Land'. The rest of Britain was a warm shallow sea. Into this sea calcium carbonate was deposited. The sea bed was gradually subsiding and so a great thickness of sediment built up – about 1000 m.

This forms the Carboniferous Limestone – a thick grey rock with many fossils which forms the distinctive type of landscape known as *karst*. Karst scenery is described in Chapter 4. Carboniferous limestone has a clear pattern of joints and bedding planes and is therefore often shown on geological diagrams with a type of shading that resembles brickwork. Water passes through the rock by seeping along the bedding planes and joints. By the middle of the Carboniferous period the seas were becoming filled as deltas grew from the land areas. These deltas were composed of sand grains. As the sands were squeezed together and cemented they formed the coarse gritty sandstone called Millstone Grit. They were so-called because they were used for grinding corn in historical times. These rocks do not allow water to pass through them.

As the deltas increased in size tropical forests grew on them. The landscape of Britain at that time must have been similar to the Amazon Basin today. By examining fossil plants such as the one in Fig. 13 we can obtain a clearer idea as to what Britain must have been like in Carboniferous times.

The sea level then rose and fell many times. With each rise of sea level the forests were drowned and mud was deposited on top. More deltas were formed and the forests that grew on them were once again drowned. This process was repeated many times. The drowned forest vegetation was changed into *peat* and then to coal as it became buried under layers of mud and sand. The bands of coal are called *seams*. The *coal measures* refer to all the rocks of the late Carboniferous period.

15 Make a tracing of Fig. 14b. Using an atlas or textbook of British Isles geography shade the main coalfield areas of Britain. Why are there no coalfields in North Scotland or central Wales?

16 Refer to the map and cross-section of the Yorkshire, Derbyshire and Nottinghamshire coalfields (Fig. 12).
 a The Permian rocks are not horizontal. In which direction do they dip?
 b To the west of the coalfield the coal measures reach the surface. To the east they are hidden by younger rocks. These two parts of the coalfield are called the *exposed* and *concealed* coalfield. Which part do you think was mined first and why? Make a copy of the map and label the exposed and concealed coalfields.

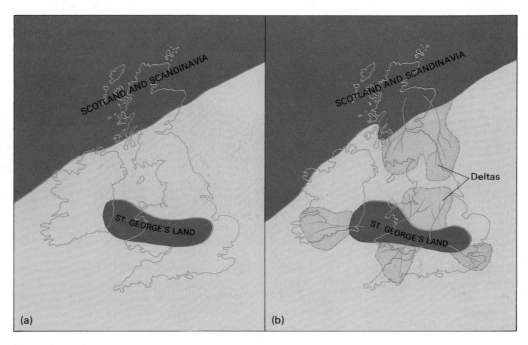

Fig 14 Britain in the Carboniferous period   a) about 340 million years ago   b) about 290 million years ago

Fig 15 How coal is formed

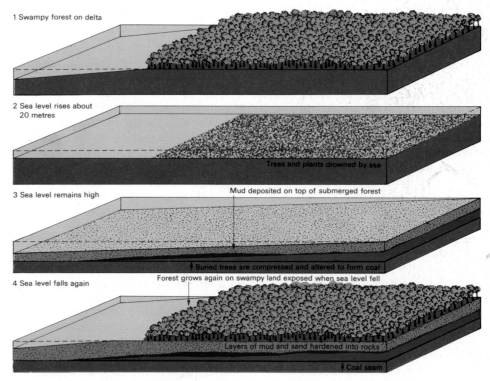

1 Swampy forest on delta

2 Sea level rises about 20 metres
Trees and plants drowned by sea

3 Sea level remains high
Mud deposited on top of submerged forest
Buried trees are compressed and altered to form coal

4 Sea level falls again
Forest grows again on swampy land exposed when sea level fell
Layers of mud and sand hardened into rocks
Coal seam

Water in rocks

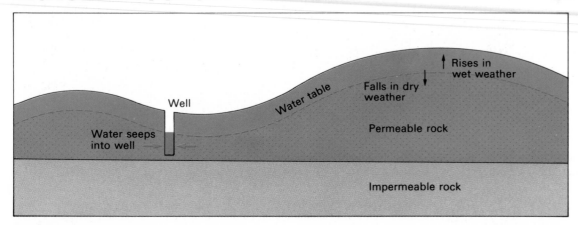

Fig 16 The water table

Rocks which allow water to pass through them are called *permeable* rocks. Those which will not let water through are *impermeable*. Permeable rocks are of two sorts, porous and pervious. *Porous* rocks have air spaces in them like a sponge. Water passes through the pores. Chalk is a porous rock.

17 Weigh a piece of dry chalk rock. Soak it in water, and then reweigh to see how much water the chalk has absorbed.

Carboniferous Limestone will not allow water to pass through the rock itself but water will seep through a layer of the rock because the rock has cracks. Water passes through the bedding planes and joints. Rocks like this are called *pervious* rocks.

Water may be stored underground depending on the arrangement of the permeable and impermeable rocks. Water will remain in a permeable layer if there is an impermeable layer beneath it. Rock strata in which water is stored are called *aquifers*. The top level of water saturation in rock is termed the *water table* although, unlike a table, it is not flat but follows the surface relief (see Fig. 16). A well is a hole deep enough to reach the water table. Water seeps in through the sides of the well until the water level is the same height as the water table. The water table may rise or fall according to, for example, the amount of rain received. In a dry spell the water table could fall below the level of the well, in which case the well would run dry.

18 Fig. 17 shows a map of the area around London served by the Thames Water Authority. Study the map and the following statistics:

Total number of people in the area
8 million
Average total quantity of water used per year
810 300 million litres
Total quantity of water pumped each year from boreholes
225 000 million litres
Total quantity of water pumped each year from rivers
650 000 million litres

a The area shown on the map used to be managed by many small separate water authorities, each taking water from wells and rivers, purifying it and returning the used water to rivers. What advantages are there in having one large authority controlling all water uses in this area? (Remember that the T.W.A. is responsible for sewage disposal, fishing and recreational use of water as well as the water supply to houses, factories and farms.)

b How much water is used per head of the population living in the area per day? Estimate how much each person might drink each day. What is the rest used for?

c Approximately what percentage of the water pumped each year comes from underground?

19 Drip water onto specimens of shale, granite, Carboniferous Limestone, Millstone Grit, chalk and clay. Which rocks allow water to pass through?

20 Study Fig. 17.
a List the major aquifers in the London Basin.
b Which upland areas will absorb water when rain falls?
c Describe in detail why the gravel is able to act as an aquifer. Would it behave in the same way if it lay directly above the chalk?
d Why are there springs in the position shown on the cross-section?

Lay the edge of a ruler across the highest points of the water table in Fig. 17. Is the top of the well above or below this line? If it is below it means that water seeping into the hole should gush out of the top of the well like a fountain under its own pressure. A saucer-shaped arrangement of the rocks like this is called an *artesian basin.*, and the well an *artesian well*. The fountains in Trafalgar Square once worked by natural artesian pressure but now the water has to be pumped up. This is because the demand for water has been so great that the pressure in the chalk aquifers has dropped.

21 In some areas within the Thames Basin water is pumped underground for storage rather than keeping it in surface reservoirs. What advantage might this method of storage have?

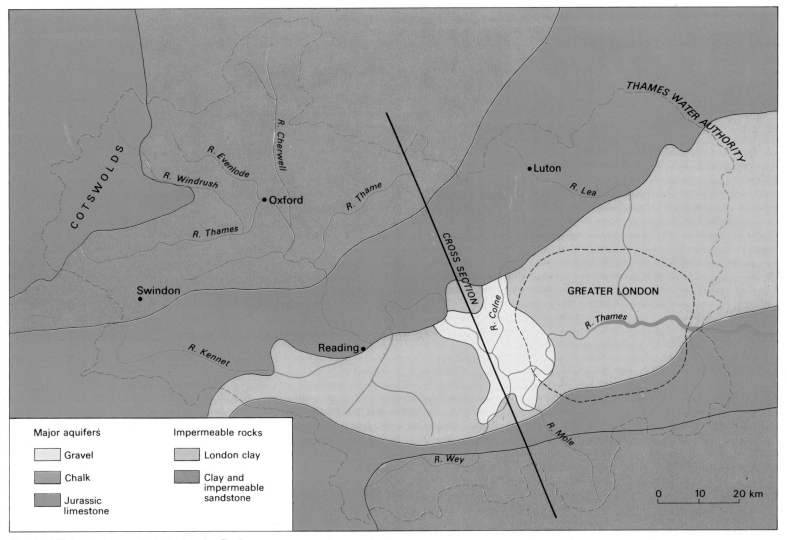

Fig 17 a) Water-bearing rocks in the London Basin
b) Cross-section of the London Basin

Major aquifers
Gravel
Chalk
Jurassic limestone

Impermeable rocks
London clay
Clay and impermeable sandstone

0   10   20 km

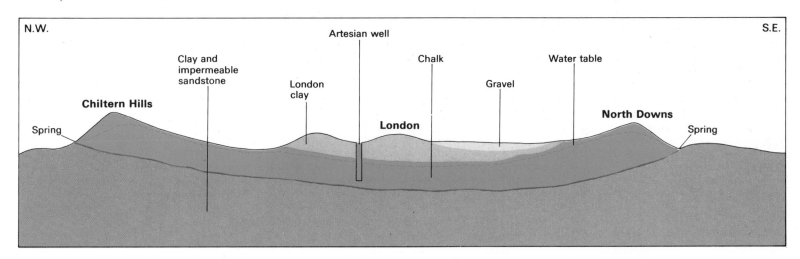

N.W.   S.E.

Clay and
impermeable
sandstone   London
clay

Artesian well

Chalk   Water table

Chiltern Hills   Gravel

Spring   **London**   **North Downs**

                                                                                            Spring

Chalk and chalk landscape

The nature of rocks, whether they are hard or soft, permeable or impermeable, affects the type of landscape. Chalk, for example, is a soft, permeable, white rock. Note that natural chalk is not the same as blackboard chalk although it is very similar in appearance.

Chalk was formed during the Cretaceous period. Tiny sea creatures called *coccoliths* – about 0.005 mm in diameter – made a kind of armour plating for themselves from calcium carbonate. Chalk is formed from the gradual accumulation of these tiny protective shells on the sea floor. The sea which covered Britain at this time must have been very clear and calm as the chalk is almost pure white calcium carbonate – only 1 per cent of it is clay.

22 Suppose the total thickness of the chalk is about 250 m. The Cretaceous period lasted about 65 million years. How many years did it take for one metre of chalk to be deposited? How long for one millimetre?

Fig 18 Early flint tool

Chalk contains flint. Flint is found in knobbly lumps, which are very hard and when broken have very sharp edges (see Fig. 18). What use did early man make of flint?

23 Fig. 20 shows the chalk scenery of the South Downs. Describe the landscape. Are the hills gentle or rugged?

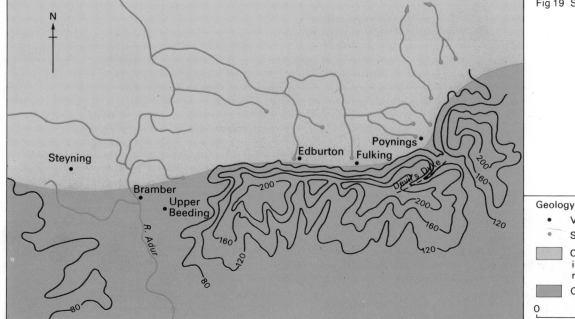

Fig 19 Springs and dry valleys in the South Downs

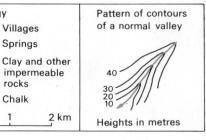

Fig 20 The South Downs near Fulking, Sussex

24 Refer to the map of the South Downs (Fig. 19).
a Describe and account for the position of the springs that appear on the map.
b Look at the area marked Devil's Dyke on the map and the photograph (Fig. 21). The pattern of contours shows a deep valley but there is no river or stream. Valleys such as this are called *dry valleys*. Using a piece of tracing paper mark the position of the other dry valleys on the map. On which rock type do they appear?

Chalk is porous and so water seeps into it. How, therefore, could streams have cut valleys and why did those valleys disappear? One suggestion is that these valleys were cut by rivers in the Ice Age. This part of Britain was not covered by ice but the ground here beneath the surface was frozen. This frozen ground would have been impermeable. Valleys could have been cut by streams flowing over it.

Review Questions

1 Consider the use that we make of rocks. Make columns like those below and list ten rocks, showing how they are put to our service.

Rock   *Use*
Slate   Formerly used for roofing. Now largely used for ornamental stone.

2 Write a brief paragraph on each of the following words: joint, dip, ore, Tees-Exe line, coal measures, aquifers.

3 Draw three columns, labelled *Igneous*, *Sedimentary* and *Metamorphic*. Put each of the rocks below into its correct column.
Millstone Grit, marble, slate, granite, coal, Old Red Sandstone, basalt, clay, chalk, shale.

Fig 21 The Devil's Dyke: a dry valley

THOUSANDS FLEE HOMES AFTER U.S. EARTHQUAKE

FRANTIC efforts were under way yesterday to drain thousands of gallons of aviation fuel from ruptured storage tanks as hundreds of aftershocks rolled through the California-Mexico border in the wake of a powerful earthquake.

The earthquake, which measured 6·4 on the Richter scale, caused damage running into millions of dollars and injured 100 people on both sides of the border, two seriously. But most of them suffered only cuts from flying glass and debris.

Thousands of people slept outdoors after the earthquake, either through fear of aftershocks or because their homes were so badly damaged.

Roads, water and gas mains buckled, roofs and walls of buildings fell in, foundations were tilted and a canal was seriously damaged.

The epicentre was just south of the border the Mexican town of Mexicali and the American towns of Calexico and El Centro. These small farming communities lie along a known fault line in the Imperial Valley area.

Buildings swayed

The effects were felt in Los Angeles and Las Vegas, 200 miles away, where tall buildings swayed.

"I've seen panic in the streets in films, but never before in real life," said Mr Frank Lope, a television reporter in El Centro.

One man who remained calm, however, was Calexico's municipal judge Robert Fox. When the courthouse began shaking, he moved a hearing out of

the building on to the front pavement.

At least 15 after-shocks ranging up to five on the Richter scale were recorded, together with hundreds of smaller tremors with intensities up to two and three.

The earthquake was the strongest to affect the American mainland since the San Fernando earthquake near Los Angeles, in 1971 which measured 6·5 and killed 65 people in widespread devastation.

By Ian Brodie in Los Angeles

Fig 1 An extract from the *Daily Telegraph*, 17 October 1979

Earthquakes

Earthquake damage

People who live in the western states of North America face the constant threat of earthquakes. When the earth's crust buckles and cracks, huge forces are released which are measured on the *Richter Scale*. A figure of 6.4 on this scale represents a moderate earthquake while figures of more than 7 indicate major earthquakes.

1 Imagine an earthquake of this size happening in Britain and affecting cities 200 miles away. If it took place at Birmingham, in which other cities would the shocks be felt?

When an earthquake measuring 8.5 on the Richter Scale shook Alaska on Good Friday 1964, strong tremors lasted for four minutes and 12 000 aftershocks were recorded in the next two months. In a few seconds over 200 000 km² of land rose or fell by a metre or more. Great cracks opened up, buildings collapsed and a large area of built-up land at Anchorage, capital of Alaska, slid 150 metres towards the sea. Huge *tsunamis* (sometimes called tidal waves) were formed and these waves travelled to all parts of the Pacific Ocean. At harbours in Alaska, ships were lifted up and dropped further inland. The waves were still six metres high when they reached Crescent City in California, 4000 km away, where twelve people were drowned.

2 Alaska is sparsely populated and only 114 people lost their lives. What sort of area would 200 000 square km cover in Great Britain and Western Europe? How much damage would be done if this area were to be suddenly raised or lowered by a metre or more?

Earthquakes are very rare in Britain so we can easily forget the enormous damage they cause in many parts of the world. In 1906 an earthquake of magnitude 8.3 led to the loss of over 700 lives at San Francisco. This city lies near to the San Andreas Fault, a huge series of cracks in the earth's crust. Movement along the fault is continuing and the longer the residents of San Francisco have to wait for the next earthquake the more severe it is likely to be. It has been estimated that if a similar 8.3 earthquake is repeated at San Francisco there would be 10 000 deaths and 40 000 hospital cases. Large sums of money are therefore being spent on efforts to forecast and even control earthquakes.

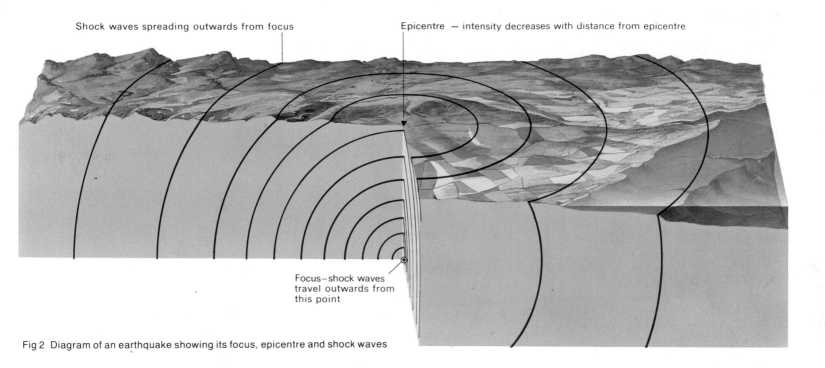

Shock waves spreading outwards from focus

Epicentre — intensity decreases with distance from epicentre

Focus—shock waves travel outwards from this point

Fig 2 Diagram of an earthquake showing its focus, epicentre and shock waves

Forecasting and controlling earthquakes

Before earthquakes can be controlled, geologists must try to understand why they happen and work out where they are most likely to occur. *Seismologists* specialise in studying earthquakes and have discovered that shocks are associated with the buckling and cracking of rocks at depths of up to 300 km.

Think what happens if you hold a plastic ruler in both hands and gradually bend it. Strains build up until eventually the ruler breaks. Energy stored up in the ruler as it was bent is released as shock waves – a cracking noise and vibrations along the broken halves of the ruler. The earth's crust behaves in a similar way. Strains build up until the crust breaks and energy is released in an earthquake.

The point of origin of an earthquake is known as its *focus* (Fig. 2). Shock waves are first felt at the *epicentre*, the place on the earth's surface above the focus, and the worst damage is usually done in this area. Several types of waves move outwards from the focus. The first waves to arrive are known as Primary (P) Waves and these are followed by Secondary (S) Waves. P and S Waves move through the earth but the last waves to arrive travel along the surface and it is these which do the most damage, causing the earth to move both up and down and from side to side. All these waves are picked up and recorded by *seismographs*.

Fig 3 Recording earthquakes: a seismograph

If the records of all the world's seismic stations are put together, it is immediately clear that some areas are greatly at risk from earthquakes. In these danger zones, the rocks are moving and cracking along *faults*. Earthquake records are used to pinpoint places along fault lines where an earthquake has not happened in recent years and where tensions must be building up. This allows seismologists to say where an earthquake is likely. To say when it will occur is far more difficult. Some progress towards successful earthquake forecasting is being made by studying changes in the properties of rocks which take place immediately before an earthquake, but the days of accurate earthquake prediction are still some way ahead. Meanwhile, scientists are experimenting with ways of controlling earthquakes. If movement along fault lines can take place as frequent, small jerks rather than a few large movements there would be less risk to life and property. Some success has been achieved by pumping water into the fault zone, which lubricates the fault and allows movement to take place before huge stresses build up. But such experiments are dangerous and are being carried out very cautiously.

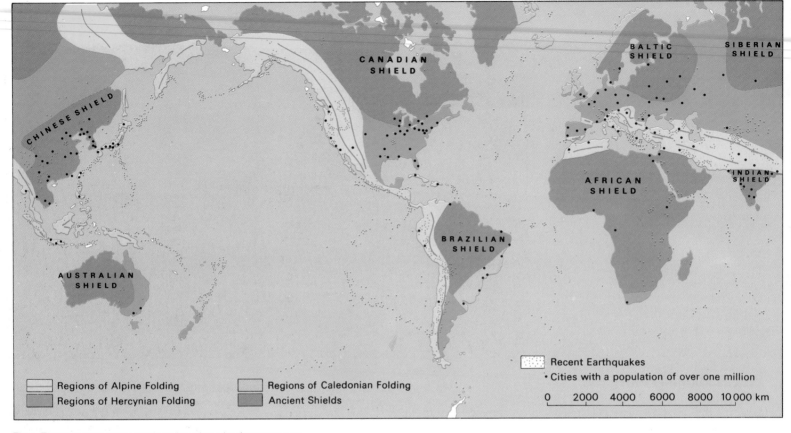

Fig 4 The structure of the earth and earthquake danger zones

3 Study Fig. 4. List ten cities with a population of one million or more which are in danger zones. You can find the names of the cities in your atlas.

Plate tectonics

The structure of the earth

We have seen that earthquakes are associated with movement along faults and that they happen most regularly in particular areas of the world. Looking at Fig. 4 we see that there are narrow zones of earthquake belts which surround large, more stable areas. This geological pattern is repeated if we study maps of major mountain ranges and volcanoes. Fig. 4 also shows the world's structural areas. The *shields* are regions of very old rocks where

there is little movement. Elsewhere mountains have formed at various periods of the earth's history. The older mountains have been worn down while the younger ones form the great mountain ranges of the world today.

4 Refer to the geological time scale on page 10. How long ago were the Caledonian, Hercynian and Alpine mountains formed?

5 On tracing paper, copy the outlines of the continents from Fig. 4. Shade the areas of Alpine fold mountains and then use your atlas to help plot the following volcanoes on the same map:

Vesuvius	41°N 14°E
Mt. Hood	45°N 122°W
Etna	38°N 15°E
Paricutin	19°N 102°W
Hekla	64°N 20°W
Santorini	36°N 25°E

Kilimanjaro	3°S 37°E
Villarrica	39°S 72°W
Cotopaxi	1°S 78°W
Ruapehu	39°S 176°E
Krakatoa	6°S 105°E
Mayon	13°N 124°E
Fuji	35°N 138°E
Lassen Peak	40°N 121°W
Mauna Loa	19°N 156°W
Mt. Pelée	15°N 61°W
Tenerife	28°N 17°W
Tristan da Cunha	37°S 12°W
Aconcagua	33°S 70°W

Place this tracing over Fig. 4. What do you notice about the location of earthquakes and young fold mountains?
How many of the volcanoes are also in these areas?

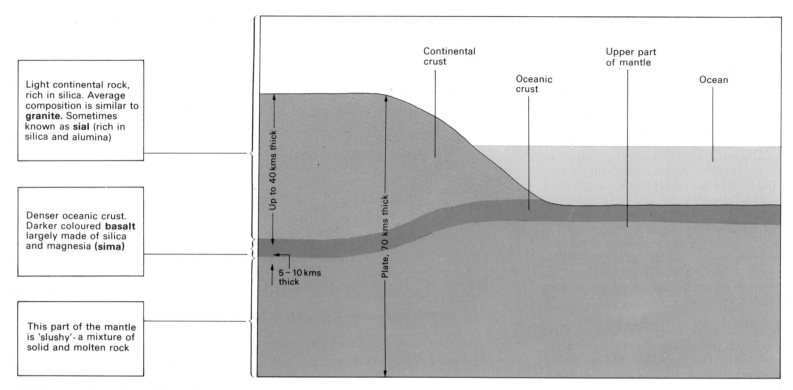

Light continental rock, rich in silica. Average composition is similar to **granite**. Sometimes known as **sial** (rich in silica and alumina)

Denser oceanic crust. Darker coloured **basalt** largely made of silica and magnesia (**sima**)

This part of the mantle is 'slushy'- a mixture of solid and molten rock

Continental crust

Oceanic crust

Upper part of mantle

Ocean

Up to 40 kms thick

5–10 kms thick

Plate, 70 kms thick

Fig 5 Cross section of the earth's crust

Most of the earthquakes and volcanoes occur in areas of young fold mountains or in the oceans where, as we shall see later, there are underwater mountain ranges as spectacular as those found on the continents. Could this be mere coincidence? In the 1960s and 70s evidence was collected which helps us explain the connexion between earthquakes, volcanoes and mountain ranges. The theory of *plate tectonics* is one of the great scientific advances of the twentieth century. It is now accepted by almost all earth scientists, though there are many areas where our understanding of the unstable earth is still incomplete. To follow this theory we first need to understand how the inside of the earth is made up.

Inside the earth

The earth has a radius of 6300 km. Its centre is very hot but is kept solid by the immense pressure of the rocks above. The *core* as this zone is called, becomes molten about 1000 km from the centre of the earth. 3000 km from the centre there is a change in composition where the core is

replaced by the *mantle*. For the most part this zone is solid and it extends almost to the surface of the earth. The very thin top layer is known as the *crust*. It is about 5 km thick under the oceans and 30 km thick under the continents – very little compared with the overall size of the earth.

6a Using a scale of 1 mm for 50 km, draw a vertical column 1 cm wide on the left hand side of your page to represent a slice from the centre to the surface of the earth. Divide the column into the solid core, the liquid core, the mantle and the crust (under the continents).

 b Alongside the column, write these notes about each layer:
   Crust: thicker under continents than under oceans
   Mantle: dense, dark coloured rocks. The upper mantle is partly molten and contains slow convection currents
   Molten core: movements here cause the earth's magnetic field. Rich in iron
   Solid core: kept solid by immense pressure from above. Rich in iron.

If we look in more detail at the crust we see (Fig. 5) that the continents are made of material which is less dense than the oceanic crust, so that they are supported by the denser material below. If the continents are worn away by erosion they become lighter still and rise but if they have to support a great weight such as an ice sheet they sink. This state of balance, known as *isostasy*, explains why much of Scotland is rising by several millimetres a year. It is gradually recovering from the great weight of ice which forced it downwards in the Ice Age.

7 Demonstrate the principle of isostasy by observing the different state of balance achieved by blocks of different densities (for example, cork, ice and various types of wood) floating in water. Measure the proportion of each block below and above the water. Which material corresponds to continental crust and which represents oceanic crust? Note that if you place a weight on one of the blocks it tends to sink and if you take the weight off, it bobs up again.

Plates on the move

According to the theory of Plate Tectonics, the upper part of the earth is divided into *plates* (Fig. 6). These are seventy kilometre thick rafts of mantle rocks and crust which move very slowly about the earth's surface, floating on the slushy part of the mantle. They move only a few centimetres a year but over millions of years this has caused continents to split apart and collide. The forces involved are immense which explains why many of the earth's most spectacular features occur at plate boundaries.

8 Compare Figs. 4 and 6. Name
a 5 'shields' which correspond to plates.
b 5 ranges of young fold mountains which are at plate boundaries. Look up the names of the mountain ranges in your atlas.

Most geologists accept the idea of Plate Tectonics though there is still uncertainty about why the plates move. One theory suggests that there are slow convection currents in the mantle, similar to those obtained if water is heated in a saucepan.

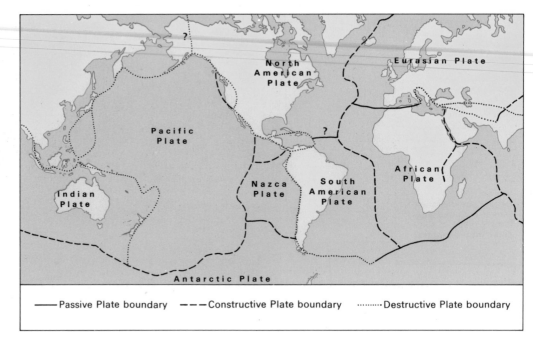

Fig 6 Plates on the surface of the earth

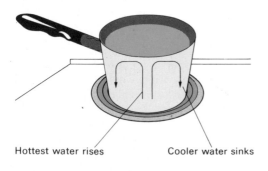

Fig 7 Convection cells in a saucepan and in the earth's mantle

Hottest water rises   Cooler water sinks

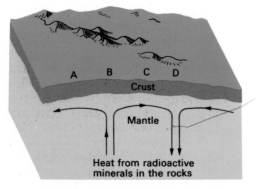

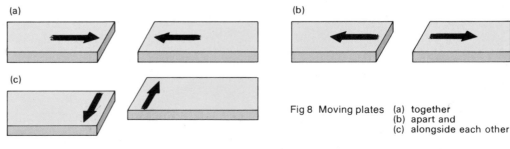

Fig 8 Moving plates (a) together
                     (b) apart and
                     (c) alongside each other

9 Fig. 7b shows convection cells in the mantle. Where on the crust (at A, B, C or D) would you expect
a the crust to split open and plates to move away from each other?
b plates to be colliding?

Plate boundaries

When plates move together and collide the plate consisting of denser, oceanic material is pushed down into the mantle where it melts and is destroyed. This is known as a *destructive plate boundary*. Study Fig. 9 and you will see that the collision of plates produces *fold mountains* and that *ocean trenches* are formed where the oceanic plate bends downwards. As the oceanic crust slides under the continent the rock is melted and some of it moves upwards. Huge *batholiths* of igneous rock are formed if this rock solidifies underground (p. 31) and some material escapes at the surface forming *volcanoes*. Contact between the two plates also produces *earthquakes*.

In other areas of the world plates are being created rather than destroyed and this most commonly happens in the centre of ocean floors. At a *constructive plate boundary* the plates are moving apart. Lava wells up to 'plug the gap' and new material is added to the outward-moving plates. *Rift valleys* (Fig. 20) show that the plates are pulling apart. The Mid Atlantic Ridge and volcanic islands along it such as Tristan da Cunha and Iceland consist of new crust forming at a constructive plate boundary.

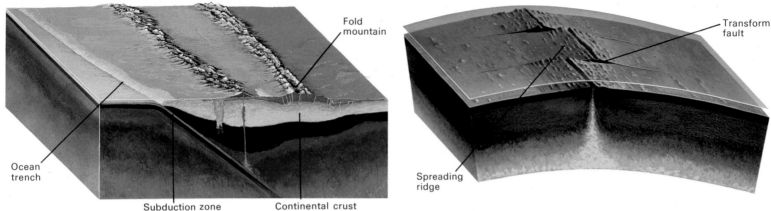

Fig 9 A destructive plate boundary

Fold mountain

Ocean trench

Subduction zone

Continental crust

Transform fault

Spreading ridge

Fig 10 A constructive plate boundary

10 Trace outlines of South America and Africa from Fig. 11. Cut round these outlines and then place them back onto Fig. 11. Push the continents together so that they meet along the Mid Atlantic Ridge with the letters A, B and C matching up. In doing this you have reversed the movements of the last 150 million years. This jigsaw puzzle fit was one of the first pieces of evidence that suggested the idea of *Continental Drift*.

Where plates are moving alongside each other, crust is neither destroyed nor created, though molten rock sometimes finds a way to the surface between the plates. If the plates were to slide smoothly past each other these *passive plate boundaries* might hardly be noticed, but the plates tend to grate against each other, causing earthquakes. We have seen how scientists are trying to control plate movement by lubricating the plate boundaries, making their movement smoother and more gradual.

11a List i) the features associated with constructive plate boundaries.
ii) the features associated with destructive plate boundaries.
b In your own words, explain how (i) Iceland (ii) the Andes were formed.
c List five volcanoes associated with constructive and five associated with destructive plate boundaries. (Use Figs 4 and 6.)

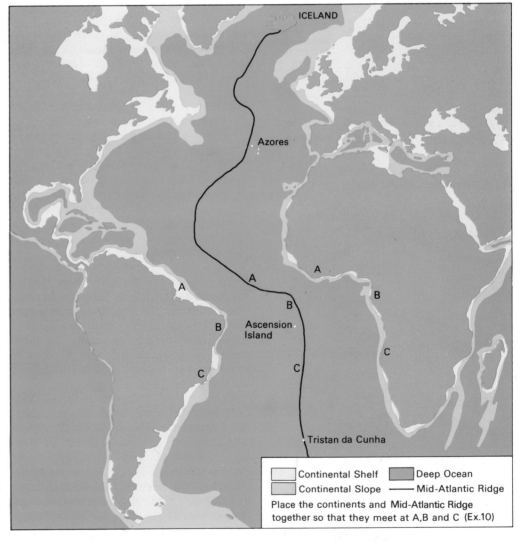

Fig 11 The Atlantic Ocean with surrounding continents and their continental shelves

23

Folding

Fig 12 An anticline

Fig 13 Anticline (A)   syncline (S)

Fig 14 Different kinds of folding

(a)

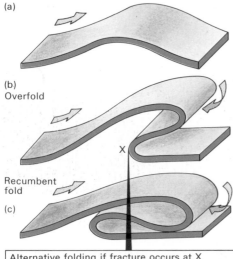

(b)
Overfold

X

Recumbent fold

(c)

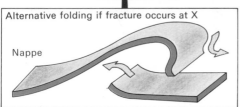

Alternative folding if fracture occurs at X

Nappe

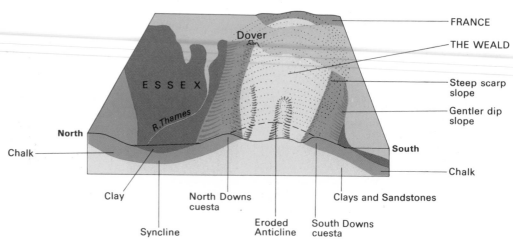

Fig 15 Folding on a large scale in S.E. England

When sedimentary rocks form, their strata are usually arranged in horizontal layers (see p. 10). Yet in many places we see these strata twisted and contorted. How has this come about?

12 Hold a sheet of paper or thin card between your hands and move your hands together allowing the paper to sag downwards. This causes a type of *fold* known as a *syncline*. Then put the paper on a table and move your hands together. The paper arches up forming an *anticline*. Match these folds with the photographs (Figs 12 and 13) and draw a cross-section of each fold in your exercise book.

Folds are usually formed when rock is squeezed at destructive plate boundaries. The results may be huge mountain ranges and valleys or tiny patterns in a piece of rock. The folds may be simple like those you have demonstrated with a piece of paper or they may be complex like those shown in Fig. 14.

Rocks in almost all parts of Britain have been folded at some stage. In South Eastern England the folds are generally simple anticlines and synclines whereas in Scotland 'recumbent' folds and 'nappes' are more common. Where folds occur at the surface they are eroded and often difficult to recognise. Few valleys are synclines and few mountains are anticlines. *Escarpments* or *cuestas* are formed by the erosion of rocks tilted by folding and these ridges

form the usual type of hill found in S.E. England (Fig. 15).

13 Write out the following paragraph, filling the gaps with words chosen from this list: cuesta, clay, chalk, south, north, scarp, dip, anticline, syncline, Weald.

The London Basin is a _____ which has been partly filled in with _____. The North and South Downs are the remains of a huge _____ which once covered the area known as the _____. **They stand up as lop-sided ridges** called _____ because they are made of resistant _____. Each ridge has a gentle _____ slope and a steeper _____ slope. The scarp slope of the North Downs faces _____ whereas the scarp slope of the South Downs faces _____.

14 Look again at Fig. 4. Notice that the youngest fold mountains in Europe are in the south – name two mountain ranges of Alpine age – while the oldest are in the north – in which countries? The same pattern is repeated in Britain. The Downs of S.E. England are Alpine, the Pennines are Hercynian and the Scottish mountains are Caledonian. Normally you would expect the youngest fold mountains to be the highest (why?) but Britain was a long way from plate boundaries in Alpine times so only small hills were formed.

Faulting

Rocks can only be bent a certain amount before the strain is too great and they fracture along a *fault* or crack. The rocks along one side of the fault may move only a few centimetres from the rocks on the other side, though movement along major faults amounts to several kilometres.

Because faults are lines of weakness in the crust they are of great practical importance in engineering. If a dam or other large structure is built above a fault, any movement may make it collapse and many lives may be lost as a result. Faults are also studied by mining engineers. Many coal seams in Britain are expensive to work because they are faulted. The most efficient mines have seams which continue at the same level for a good distance, but if the seams are crossed by faults, this will not be the case.

When large vertical movements occur along a fault, the land on one side may be raised several metres above the land on the other side. If such movements are frequent the difference in height may amount to tens or hundreds of metres. The steep slope that this produces is called a *fault scarp*. Because it is so long since movement took place along British faults, fault scarps in this country have been destroyed by erosion. Yet faults still leave their mark on the landscape. Hard and soft rocks may be brought together by faulting and when the softer rock is eroded a feature similar to a fault scarp is produced.

15 Using Fig. 18, describe the appearance of Giggleswick Scar and explain how it has been formed.

A fault is often surrounded by a zone of crushed rock which, being weaker, is easily picked out by agents of erosion. The Great Glen in Scotland follows the line of a *tear fault* (Fig. 16). This weakened the rock which was then eroded by rivers and glaciers. How much movement has taken place along the fault is not certain but you can examine one theory in the next exercise.

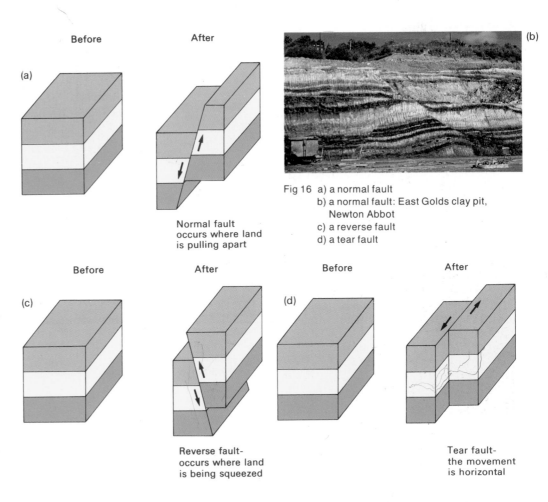

Normal fault occurs where land is pulling apart

Reverse fault- occurs where land is being squeezed

Tear fault- the movement is horizontal

Fig 16 a) a normal fault
b) a normal fault: East Golds clay pit, Newton Abbot
c) a reverse fault
d) a tear fault

Fig 18   A fault scarp: Giggleswick Scar, Settle, North Yorkshire

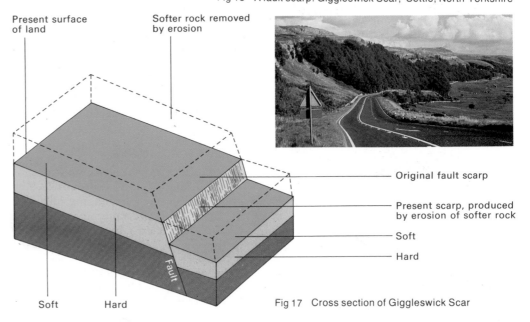

Present surface of land

Softer rock removed by erosion

Original fault scarp

Present scarp, produced by erosion of softer rock

Soft

Hard

Soft   Hard

Fig 17   Cross section of Giggleswick Scar

16 Study the map of the Great Glen. It is thought that the masses of granite at Foyers and Strontian are halves of what was once a single body of granite. Trace this map onto a piece of paper, then cut along the Great Glen Fault and slide the northern part of the map back along the fault until the two masses of granite lie together. How far have you moved northern Scotland? Remember that the movement you have just reversed took place very gradually and would have been accompanied by a series of spectacular earthquakes.

When several faults occur in the same area, earth movements may raise or lower the land between them. If the land sinks between two parallel faults a *rift valley* is formed (a rift is a crack or fault).

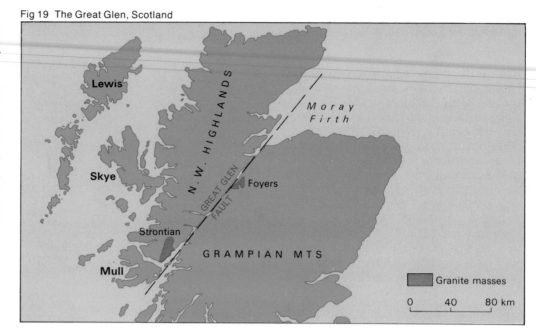

Fig 19 The Great Glen, Scotland

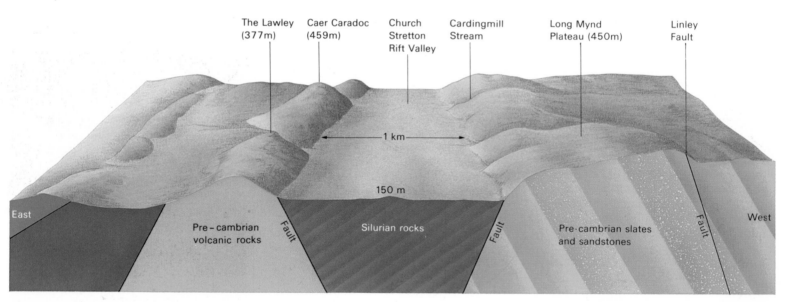

Fig 20 The Church Stretton rift valley

17 Use Fig. 20 to complete this description of the Church Stretton rift valley:

The Church Stretton rift valley is situated 20 km south of Shrewsbury in the Welsh border country. It lies between two ranges of hills, the ____to the west and the ____to the east. The rocks in the valley are *younger/older* than those of the surrounding hills.

The valley is ____wide and its sides rise from ____ m above sea level to ____m in the surrounding hills. The sides of the rift valley have been cut into ('dissected') by streams such as the ____.

If the area between two faults is raised above the surrounding land, a *horst* or *block mountain* results. The Long Mynd rising between the Church Stretton and Linley faults is an example of a horst.

18 Draw two cross sections to explain how a rift valley is formed, showing the appearance of the land before and after faulting. Then draw two similar diagrams showing how a horst is formed.

26

Vulcanicity

Volcanic landforms result from the cooling of *magma*. If this happens at the surface *volcanic* or *extrusive* rocks are formed. If it happens beneath the surface *plutonic* or *intrusive* rocks are formed.

Volcanoes

On the island of Sicily stands Etna, Europe's highest volcano at over 3300 m. It has two *craters*, but it was not from either of these that the spectacular 1971 eruption occurred. Towards the end of March a patch of snow south of the main crater began to melt and it was here that a crack opened up on April 5. Showers of molten *lava* were flung out, building *cinder cones* which were soon 30 m high. Lava flowed down the mountain, permanently ruining the best ski slopes, demolishing ski lifts and damaging an observatory built in the 1930s for studying volcanic activity. At night the red glow of the lava was reflected from clouds of *gas* and *steam*. On May 4 a new *fracture* opened and yellow-hot lava was thrown out from the *vent* building up a cone at the rate of a metre an hour. Three days later a kilometre-long fissure opened up and lava streamed out, flowing at two metres a second. On May 12 yet another crack appeared and lava poured downhill destroying forests and vineyards, engulfing a farmhouse and forcing the evacuation of the village of Fornazzo. On May 18 a 200 m wide crack appeared. Dark *ash* and white steam were flung into the air forming cauliflower-shaped clouds before falling to earth and blanketing the surrounding countryside. On June 9, gases escaping from the fissure caused an *explosion* which produced a 6 m wide crater.

19 List the different materials thrown out by Etna during its eruption.

20 Study Fig. 21.
  a Draw a cross-section of Etna from X to Y, using a vertical scale of 1 cm : 2000 m and a horizontal scale of 1 cm : 5 km.
  b How far from the fissure which opened on May 12 did lava run and how many

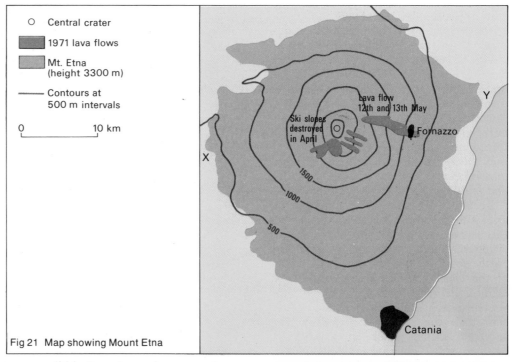

- ○ Central crater
- ▨ 1971 lava flows
- ▨ Mt. Etna (height 3300 m)
- —— Contours at 500 m intervals

0   10 km

Ski slopes destroyed in April

Lava flow 12th and 13th May

Fornazzo

X

Y

1500
1000
500

Catania

Fig 21 Map showing Mount Etna

metres did it drop down the mountainside?

This eruption did an immense amount of damage though no lives were lost. Very little can be done to control such events, though lava flows have sometimes been diverted away from villages. Volcanic eruptions are difficult to forecast, but careful measurements have revealed that the ground often swells before an eruption and several successful predictions have been based on this. One difficulty is that volcanoes may be inactive for hundreds or thousands of years before becoming active once more, catching everyone by surprise. The division of volcanoes into those which are *active*, *dormant* or *extinct* is therefore only approximate. If there has been an eruption recently and another one is likely to occur, a volcano may be described as active. If it has erupted within historic times but not recently it is known as dormant, while if there is no record of an eruption it is known as extinct.

Until methods of prediction improve, we must be content with finding out more about volcanoes – why they occur, why they have different shapes and why some eruptions are more destructive than others.

Fig 22 Mount Etna erupting

Why do volcanoes occur?

They are found where magma escapes through the crust along a line of weakness (a *fracture*) or at a single point of weakness (a *vent*). If a bottle of fizzy drink is shaken up and the stopper released, the gas in the drink forces the liquid out of the bottle, sometimes to considerable heights. Similarly, magma under the earth's surface is forced out, sometimes explosively, if it can find a way up through a fissure. This is most likely to happen in areas of folding and faulting, that is at plate boundaries.

Britain is far from active plate boundaries and therefore has no volcanoes but this was not always so. The last volcanoes erupted 50 million years ago and have been partly destroyed by erosion, though parts of the cones can still be recognised. At Edinburgh, hard rock which blocked the vent proved much more resistant to erosion than the rest of the volcano and survived as Castle Rock, an example of a *volcanic plug*. The former cone of this volcano consisted, like Etna, of both volcanic ash and lava, though it was only about a thousand metres high. These layers are preserved in the hillside of Arthur's Seat and have allowed geologists to give an impression of the original appearance of the volcano.

21 Refer back to Figs 4 and 6. There are so many volcanoes around the edge of the Pacific Ocean that this has been called the 'Pacific Ring of Fire'. With which type of plate boundary are these volcanoes associated? Name five examples of volcanoes in this Ring.

Why are some eruptions more destructive than others?

This depends on how *gaseous* and *viscous* the lava is.

A comparison of Fig. 24 (a) and (b) shows that the pressure of gases in the magma chamber is an important factor in producing a powerful eruption. If the lava is viscous – thick and treacly – it does not flow easily out of the vent and tends to solidify there, forming a volcanic plug. This results in high gas pressures and explosive volcanoes. If the lava flows more smoothly, eruptions may well occur more frequently because the lava pours more

Fig 23 The Edinburgh volcano. Castle Rock is a volcanic plug.

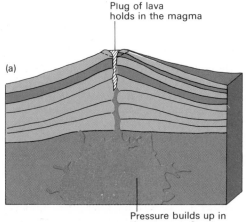

(a) Plug of lava holds in the magma

Pressure builds up in the magma chamber

(c) Vent is not blocked. Lava flows quietly out of the vent.

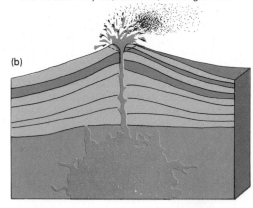

(b) Pressure blows out the lava plug throwing ash and lava out of the vent. The top of the volcano may be blown off, forming a crater

Fig 24 Magma is forced out of the earth like champagne out of a bottle

easily from the vent as soon as pressure builds up, but the eruptions are likely to be less spectacular.

Eruptions vary from *geysers*, *fumaroles* and *solfataras* which harm nobody, through *lava flows* and *explosive eruptions* to the destructive *nuées ardentes* (*fiery clouds*).

Fig 25 A fumarole. Notice the deposits of yellow sulphur

Geysers are formed when water comes into contact with hot rocks below the surface. It is converted into steam which bursts up through a fissure forming a jet which may be metres or even hundreds of metres high. Fumaroles and solfataras are small springs or vents giving off sulphurous gases. They are often all that remains when a volcano is approaching the end of its active life.

At the other extreme, it was a fiery cloud of white hot ash and gas that rushed down from Mt. Pelée at 150 km/hr to annihilate the population of St. Pierre below. In two minutes on 8 May 1902, over 30 000 people died, water in the harbour boiled and the city was destroyed. There were only two survivors.

22 Consult library books and find out more about a famous eruption such as Mt. Pelée (1902), Krakatoa (1883) or Vesuvius (A.D. 79).

Why do volcanoes have many different shapes?

Types of volcano	How it forms
a Acid lava domes and spines Example; Mt. Pelee	Acid lava is viscous and does not flow easily. It soon solidifies, building up a steep cone near the vent. Sometimes the lava is so viscous that it forms a spine sticking up from the vent rather than flowing away down the sides of the volcano.
b Basic lava shield volcanoes Example; Mauna Loa	Basic lava flows more easily and spreads over a wide area, forming a volcano with gentle slopes. Outpourings of lava may be frequent but are not usually accompanied by violent explosions.
c Composite cones or 'strato volcanoes' Example; Mt. Etna	Eruptions are sometimes explosive, emitting ash and stones, and sometimes more gentle. The volcano therefore has layers of both ash and lava.
d Ash and cinder cones Example; Mt. Paricutin	Layers of ash and cinders, varying in size, build up a symmetrical cone with a large crater.
e Calderas Example; Crater Lake, Oregon	After a major eruption the magma chamber has been largely emptied, and the volcano sinks, leaving a huge crater. Calderas (craters more than 1 km across) may also be caused by violent explosions.

Examples

a *Mt. Pelée* (Martinique, W. Indies) developed a dome within its crater before the catastrophic eruption of 1902. Later the spine shown in the photograph was forced through the dome though it crumbled within a year. This type of volcano is often associated with destructive eruptions and glowing clouds.

b *Mauna Loa* in the Hawaiian Islands is 10 000 m high, of which 4000 m are above sea level. The base of the volcano has a diameter of 400 km. During eruptions lava may flow up to 50 km down its gentle (2°–10°) slopes.

Spine——

Fig 26 Mount Pelée, Martinique, West Indies

Fig 27 A shield volcano, such as Mauna Loa

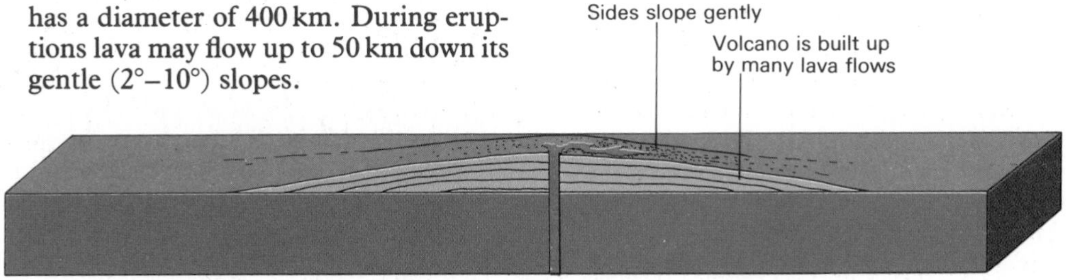

Sides slope gently

Volcano is built up by many lava flows

c *Etna* has a larger proportion of lava than ash. The two types of eruption may occur at separate times or together, as they did in 1971. See Fig. 21 for the dimensions of the volcano.

d *Paricutin* in Mexico started as a small crack in a cornfield on February 20, 1943. Within a week it was over 100 m high. Lava flowed from the base of the cone, which consists of cinders and solidified lava bombs sloping at 30°.

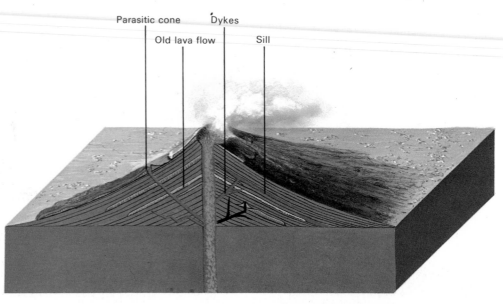

Fig 28 A composite cone, such as Mount Etna

Fig 29 Paricutin, Mexico

Fig 30 Crater Lake and Wizard Island, Oregon, USA

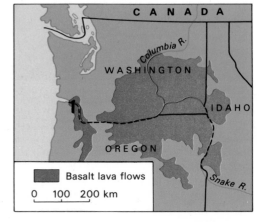

Fig 31 The Columbia Snake Plateau, USA

e *Crater Lake* in Oregon, U.S.A. is roughly circular and about 9 km across. The base of the caldera is 600 m below and the surrounding walls 600 m above lake level. The caldera was formed in a series of violent eruptions about 6000 years ago.

23 Which type of lava is associated with violent explosions and which is associated with more gentle eruptions?

24 Explain why basic lava results in wide, gently sloping volcanoes.

25 Draw a series of diagrams showing the birth of a volcano as a cinder cone, its later development into a composite cone and its destruction when a caldera is formed.

26 From the information given about Mauna Loa, draw a cross-section using a horizontal scale of 1 cm : 20 km and a vertical scale of 1 cm : 8000 m. Compare this with your cross-section of Etna (Exercise 20), remembering that this was drawn with horizontal and vertical scales four times larger. Explain the differences you note.

Basalt plateaus

When runny basaltic lava is erupted through fissures rather than a single vent, it flows downhill and starts to fill up the valleys. When this is repeated time after time valleys and then hills are covered over. The plateau this forms may spread over vast areas with little variation in height except where it has been dissected by rivers. Fig. 31 shows the 130 000 km² covered by the Columbia-Snake Plateau in the U.S.A. where hills 1500 m high were covered by hundreds of separate lava flows.

The Antrim Plateau of Northern Ireland was formed in the same way. It is part of a plateau whose other remains are found in the Hebrides, the Faroe Islands, Iceland and Greenland. The Giant's Causeway in Antrim (Fig. 32) clearly shows the joints which formed when the basalt cooled and shrank.

Intrusive features

Only a small proportion of the magma which forces its way upwards through the crust eventually solidifies on the earth's surface. Most cools within the crust, *intruded* between existing rocks. Intrusions vary greatly in size and depth. *Batholiths* are giant intrusions, deep underground. In Devon and Cornwall parts of a batholith have been exposed at the surface by the erosion of overlying rocks but the main intrusion lies 12 km below. As well as solidifying in such huge masses, magma may cool to form thinner sheets of rock. Where the land is pulling apart, fissures are filled with igneous rocks forming *dykes* which cut across the existing strata. When exposed at the surface these may be recognised as troughs or ridges depending on their resistance to erosion. Sometimes the magma follows the strata and solidifies to form a more or less horizontal *sill* or blisters up into laccoliths (Figs 35, 36)

Fig 32 The Giant's Causeway, Co. Antrim, N. Ireland

Fig 33 A batholith beneath Devon and Cornwall

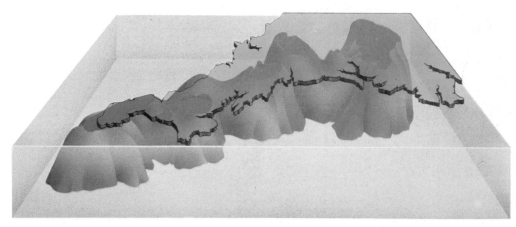

Fig 34 A dyke, Isle of Arran, Scotland

The Whin Sill is composed of dolerite, a coarse basalt which was intruded into the Carboniferous rocks of the Pennines. Erosion has revealed it in several places and it forms a prominent slope followed for several miles by Hadrian's Wall.

The Cleveland Dyke consists of basaltic rock which was intruded as a long, narrow band of rock during Tertiary earth movements. It has been extensively quarried, particularly for use in road building. The dyke can be traced from Yorkshire to the Hebrides.

Fig 35 The Whin Sill

31

Traprain Law is a laccolith which has been uncovered by the erosion of overlying rocks. It stands 150 m above the surrounding countryside, 30 km east of Edinburgh. Its steep slopes and thin soils developed on resistant rock make it unsuitable for cultivation.

27 Study the intrusive features shown in Fig. 37. Draw a simple cross-section to illustrate each of the following: sill, dyke, laccolith.

Fig 36 Traprain Law

Fig 37 Intrusions of magma beneath the surface

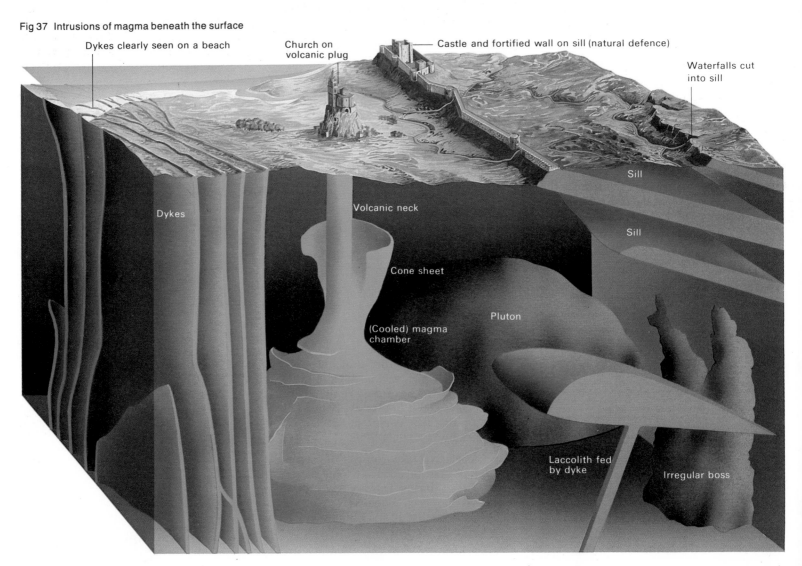

Dykes clearly seen on a beach

Church on volcanic plug

Castle and fortified wall on sill (natural defence)

Waterfalls cut into sill

Sill

Dykes

Volcanic neck

Sill

Cone sheet

Pluton

(Cooled) magma chamber

Laccolith fed by dyke

Irregular boss

Volcanoes and people

Despite the obvious danger, many volcanic areas are densely settled. Several major towns lie under the shadow of volcanoes – Naples near to Vesuvius for example – but statistics confirm the instincts of the townspeople that disaster is unlikely. On average there are 800 victims of volcanic eruptions a year but most of these people are drowned by tsunamis and may live at some distance from the eruption. In one of the worst disasters, the destruction of St. Pierre by 'fiery clouds' in 1902, the town could have been evacuated in good time if officials had not persuaded the townspeople to remain so as to take part in forthcoming elections. Statistically the risks from volcanoes, even for people living nearby, are less than the risks of disease, road accidents, famines and wars.

Vulcanicity has brought many benefits to us. When volcanic ash and lava are weathered they often produce fertile soil which will support a large agricultural population. The coffee growing lands of Brazil and the cotton growing Deccan plateau of India are based on volcanic soils. Metals of volcanic origin include mercury, silver, gold, lead, uranium, iron, copper, nickel, manganese and titanium. Steam from volcanoes and hot areas of the crust is used to generate electricity in Tuscany (N. Italy), California, New Zealand and Mexico. Hot water and steam in Iceland are used to provide heating for homes and greenhouses. Volcanic regions are also popular with tourists. The crater of Vesuvius, the glowing lavas of Etna, the geysers of Iceland and the Yellowstone National Park (U.S.A.) are visited by thousands every year. But there is still a risk: in the summer of 1979 a coachload of tourists viewing Etna at close quarters was surrounded by lava and several people lost their lives.

The spectacular eruption of Mt. St. Helens (Washington State, U.S.A.) in May 1980 illustrates some of the problems caused by vulcanicity. Warnings had been given that an eruption was expected so most of the local people had left the area. Even so, a nuée ardente (p. 29) caused some loss of life. More than a cubic kilometre of rock disappeared from the volcano, most of it pouring along a nearby valley. Some of this material was washed into the Columbia River which had to be dredged at great public expense. The fine ash was carried into the atmosphere and then drifted eastwards causing inconvenience and expense to many millions of Americans when it landed on streets and houses, damaged car engines and destroyed crops.

28 Use newspapers to keep a record of earthquakes and volcanic eruptions during the next year. Plot their locations on a world map and note the amount of damage they cause.

Fig 38 The eruption of Mt. St. Helens, Washington, USA

Review Questions

1 Explain the meaning of the following terms:
   tsunami, epicentre, geyser, vent.
2 What is the 'Pacific Ring of Fire'? Why does it occur?
3 How has the Mid Atlantic Ridge been formed?
4 How are scientists attempting to control earthquakes?
5 Name an example of each of the following:
   a rift valley
   a cinder cone
   a shield volcano
   a sill
6 Write a description of what happens in either a major earthquake or a large volcanic eruption. You may prefer to write the account as if you were actually there.
7 Draw labelled diagrams to show what is meant by
   a destructive plate boundary
   a composite cone
   a batholith

4 Weathering and slopes

Fig 1 Weathered rock

1 Look at the photograph on the left. Describe the surface of the gravestone. What has happened to the rock? What changes have taken place on the surface?

2 Look carefully at as many outcrops of rock as you can find. You may live in an area where rock appears near the surface but, if not, look at stone walls, buildings or gravestones. Describe in detail the surface of the rock. Is it crumbling or splitting? Has the surface changed colour? Can you explain these changes? You may be able to calculate the rate of rock decay by, for example, noting the date of the building or gravestone.

Weathering is the term given to the breaking down of rocks at or near the surface of the earth. This word is used because it is the weather that is largely responsible.

Rocks within the crust are at a higher pressure and temperature than those at the surface. They are stable. As the surface of the earth is gradually eroded the rocks become exposed to the temperatures and moisture of the atmosphere. There is also less pressure on the rock when it is exposed at the surface. As a result, rocks become more unstable and break up.

Rocks may be weathered in many ways but it is useful to group the processes into three types – physical, chemical and biotic weathering.

Physical weathering

Physical weathering is the splitting of rocks by stress and strain.

3 Study carefully the annotated diagram of physical weathering in Fig. 2.

a Write a brief paragraph on each of the types of weathering. Give each paragraph a heading.

b Which of these processes rely on pressure to assist in the breakdown of rock?

c How might the climate of an area determine the type and extent of weathering that takes place?

Fig 2 Physical weathering

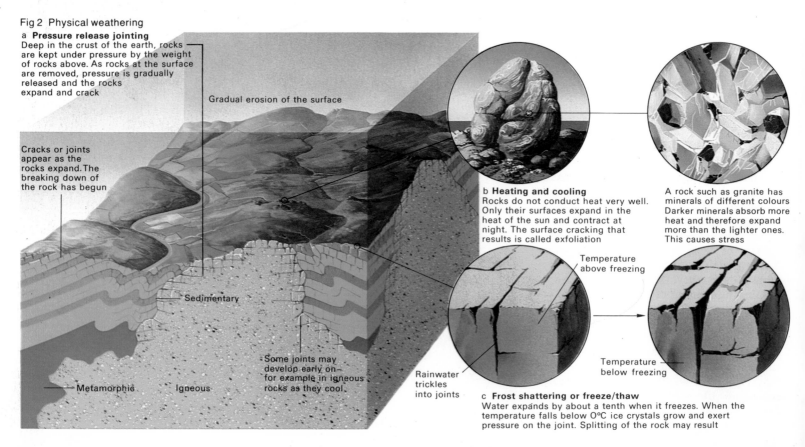

a **Pressure release jointing**
Deep in the crust of the earth, rocks are kept under pressure by the weight of rocks above. As rocks at the surface are removed, pressure is gradually released and the rocks expand and crack

Gradual erosion of the surface

Cracks or joints appear as the rocks expand. The breaking down of the rock has begun

Sedimentary

Metamorphic   Igneous

Some joints may develop early on – for example in igneous rocks as they cool

Rainwater trickles into joints

b **Heating and cooling**
Rocks do not conduct heat very well. Only their surfaces expand in the heat of the sun and contract at night. The surface cracking that results is called exfoliation

A rock such as granite has minerals of different colours. Darker minerals absorb more heat and therefore expand more than the lighter ones. This causes stress

Temperature above freezing

Temperature below freezing

c **Frost shattering or freeze/thaw**
Water expands by about a tenth when it freezes. When the temperature falls below 0°C ice crystals grow and exert pressure on the joint. Splitting of the rock may result

34

Chemical weathering

Chemical weathering is the breakdown of rock as a result of chemical reactions, usually involving rain water.

When iron is exposed to the air and moisture it turns red. This rust shows that a chemical change has taken place. Some rocks contain iron and when exposed to the air they too become 'rusty'.

There are other types of chemical weathering too. Pure rain water can react with many of the minerals that form rocks and gradually dissolve them. Water can also cause some minerals to expand, setting up stresses within the rock in the same way as heating by the sun does.

Rain water, however, is rarely pure. As it passes through the air it absorbs carbon dioxide and becomes a weak acid called carbonic acid. More carbon dioxide is added as the rain water passes through the soil (see Fig. 3). This acid reacts with or dissolves some rock forming minerals, e.g. felspar. As granite contains much felspar it is a rock that is steadily rotted by rain water.

Carbonic acid also breaks down calcium carbonate (of which limestone is composed) into calcium bicarbonate. This is soluble and is therefore carried away by rain water.

$$CaCO_3 + H_2CO_3 \rightarrow Ca(HCO_3)_2$$
Calcium Carbonate + Carbonic Acid
$\rightarrow$ soluble Calcium Bicarbonate

It is this Calcium Bicarbonate, dissolved in the water of limestone areas that makes the water 'hard'. It may later be deposited as 'fur' on the inside of pipes and kettles.

This type of weathering is called *carbonation* and its effect on limestone produces a distinctive type of scenery known as *karst* scenery (after the name of the region in Yugoslavia where it is well developed). Karst scenery is described later in this chapter.

Chemical weathering is especially effective in urban and industrial areas. Air pollution helps rain water to attack the stonework of buildings in towns. Many buildings in our cities are now being cleaned of the blackened surfaces they gained in the nineteenth century when there were no 'smokeless zone' laws.

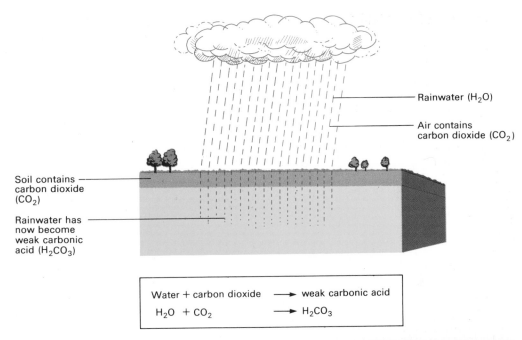

Soil contains carbon dioxide (CO_2)

Rainwater has now become weak carbonic acid (H_2CO_3)

Rainwater (H_2O)

Air contains carbon dioxide (CO_2)

| Water + carbon dioxide | $\rightarrow$ | weak carbonic acid |
| H_2O + CO_2 | $\rightarrow$ | H_2CO_3 |

Fig 3 Chemical weathering by rain water

Biotic weathering

This type of weathering is a mixture of physical and chemical weathering, caused by plants and animals.

4 Look at the photographs on the right.
a Describe how plants and animals can assist in the breaking up of rock.
b Vegetation cover on the surface of the earth provides shade. Plant roots retain moisture. How do you think these two characteristics of vegetation affect the weathering process?
c How do plants make chemical weathering more effective?

Another type of biotic weathering is found in coastal areas. Some animals, such as limpets secrete a type of acid which may attack rock.

5 Make a list of the types of weathering processes that rely on water for their effectiveness.

Fig 4 Biotic weathering: plants

Fig 5 Biotic weathering: animals

35

Climate and weathering

Because so many types of weathering rely on water and on temperature changes you will see that climate can play an important part in determining the type and rate of weathering. Many chemical reactions take place more rapidly at higher temperatures, so weathering in hot wet tropical climates is especially effective. Chemical weathering does not take place so rapidly in desert regions because of the lack of moisture. However, the small amount of water that there is in deserts plays an important part in rock decay.

6 Fig. 6c shows a feature known as a *tor* on Dartmoor in south-west England. Describe the tor, noting its size and shape.

It is not known for certain how these features were formed but it is thought they are the result of deep weathering below the surface. Some geomorphologists think this happened at a time when the climate of Britain was warmer, others believe the climate was colder than at present. Fig. 6a shows how the joints of the granite are widely spaced in some parts and more closely spaced in others. The closely jointed areas are weathered more deeply. The waste was removed leaving the areas in between rounded but standing up as tors.

Weathered material

The surface layers of rock fragments help to form part of the soil, which also consists of plant and animal matter. The type of soil varies from one place to another because it has formed from different underlying rock, in areas with different climate and vegetation, on slopes of varying angles and over different periods of time. The layer of soil and rock waste (the *regolith*) on the surface is very important to the weathering process as it holds water within it – encouraging chemical weathering. Some rock though, like limestone, has a very shallow soil as there is little of the rock left after the calcium carbonate has been dissolved and washed away.

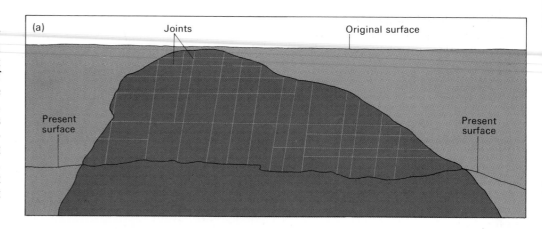

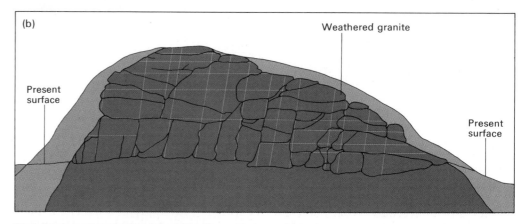

Fig 6 The formation of a tor (above). Hay Tor on Dartmoor (below)

7 Complete the following diagram, adding labels to the arrows to show the various factors that affect the type of soil formed.

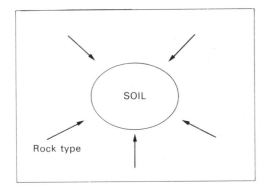

The speed of weathering

8 Look at the gravestone in Fig. 1. Estimate how much of the surface has been weathered. The stone is dated 1892. Work out the thickness of rock that has been removed each year.

It is difficult to descibe accurately the speed of weathering as it varies from place to place. It has been estimated though that 0.2 mm of rock has been removed from the surface of the pyramids of Egypt every year since they were built, and that a total of 1 cm has been removed from the limestone of St. Paul's cathedral in the last 250 years.

9 Look again at the types of weathering shown in Fig. 2. You should now be able to explain what processes have affected the rocks in each picture.
Refer back to your answer to Exercise 1. Give the photograph a more scientific label.

Limestone

Chemical weathering is especially effective on limestone. Some types of limestone develop a number of distinctive landforms which together make up a type of landscape known as *karst scenery*.

10 Look at Fig. 7 which shows a map of the Ingleborough area of North Yorkshire, and Fig. 8, the photograph of Ingleborough Hill. What do you notice about the streams in the area around Ingleborough Hill?

a Find Ingleborough on an atlas map of the British Isles. Then make a tracing of Fig. 8. Show all the streams and rivers and the contours. Label Ingleborough Hill, Chapel-le-Dale and Clapham.
Your tracing shows that streams in this area begin as springs and then apparently disappear. Some of them reappear further downslope.

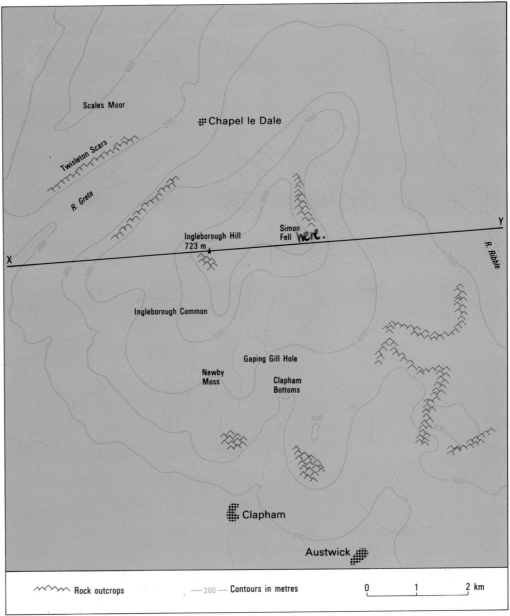

Fig 7 The Ingleborough area, North Yorkshire

Fig 8 Ingleborough Hill

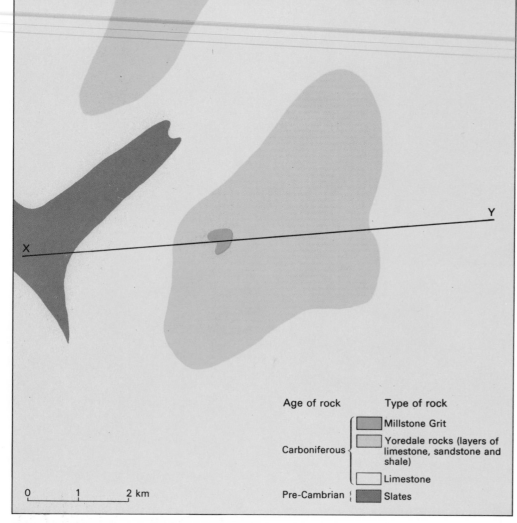

Age of rock | Type of rock

Carboniferous {
- Millstone Grit
- Yoredale rocks (layers of limestone, sandstone and shale)
- Limestone

Pre-Cambrian | Slates

0   1   2 km

Fig 9 a) Geology of the Ingleborough area (above) b) Cross-section of Ingleborough (below)

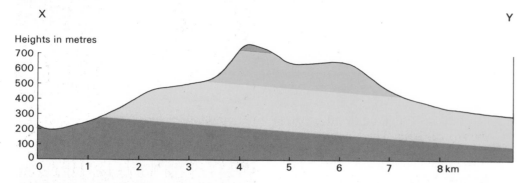

X   Y

Heights in metres

700
600
500
400
300
200
100
0
   0   1   2   3   4   5   6   7   8 km

b Place your tracing over the geological map of Ingleborough (Fig. 9a). Trace the outline of the rock outcrops and lightly shade each one. Give your tracing a key to the types of rock shown.

c Draw a cross-section along the line X–Y and label Ingleborough Hill.
Mark on your cross-section where the rock type changes and join these below the surface of the ground. Your cross-

section should look like Fig. 9b. The rocks are almost horizontally-bedded, dipping only slightly to the east. Shade your cross-section using the same colours as your map.

d On which type of rock do the streams seem to disappear? Label the cross-section with arrows to show where the streams disappear and re-appear.

e By the side of the key to your tracing label the Millstone Grit and the Carboniferous Limestone, saying whether they are permeable or impermeable. The 'Yoredale Rocks' are layers of limestone, sandstone and shales. From your tracing do you think they are permeable or impermeable?

Fig 10 The horizontal bedding planes and vertical joints can be seen in this limestone cliff

Fig. 10 shows the characteristic well-jointed and 'blocky' appearance of Carboniferous Limestone. Notice how the limestone has split along the joints and bedding planes. Although there are many different types of limestone it is mostly the Carboniferous Limestone that forms the type of landscape known as *karst scenery*.

Carboniferous limestone is made up largely of calcium carbonate, which is easily dissolved by rain water. There is

Fig 11 Limestone pavement near Horton-in-Ribblesdale, North Yorkshire

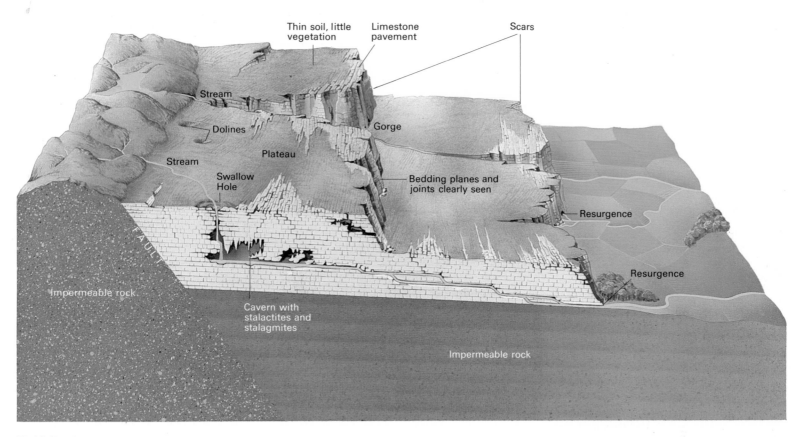

Fig 12 Karst scenery

Fig 13 Gaping Gill, near Ingleborough

therefore very little else left to form soil, so soils of limestone areas are thin, as can be seen from the photograph of Ingleborough (Fig. 8). Fig. 11 shows a *limestone pavement* where there is virtually no soil at all. The soil of this area was scraped away by the ice sheets during the 'Ice Age' and as soil on limestone forms so slowly much of this area is bare rock today.

11a Using the photograph of the limestone pavement (Fig. 11) describe the surface of the pavement.
What are the similarities and differences between this natural feature and a man-made pavement?
  b Make a detailed drawing of part of the pavement surface. Label the ridges of limestone '*clints*' and the gaps in between '*grykes*'.

  c Carboniferous Limestone weathers into large blocks separated by joints. Rain water dissolves calcium carbonate. How is the pattern of clints and grykes in the pavement formed?

Carboniferous Limestone is permeable. Water passes through the many joints in the rock, not through the rock itself. The Figure above shows how streams flowing from an area of impermeable rock onto Carboniferous Limestone may disappear down enlarged joints or *swallow holes*. These streams continue to flow underground and generally re-appear at a *spring* or *resurgence*. Some swallow holes are very deep. Fell Beck near Ingleborough (Fig. 13) falls 110 m into Gaping Gill – making one of the largest waterfalls in Britain.

Around Gaping Gill there are a number of funnel-shaped depressions or *dolines*. You can see below how these are formed. As the limestone is dissolved surface material subsides and is washed into the funnel like sand in an egg-timer.

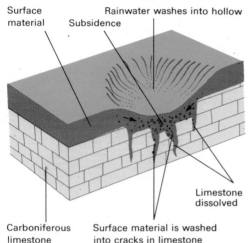

Fig 14 The formation of a doline

Some of the best-known limestone landforms occur between swallow holes and stream resurgences – underground. As running water passes through the limestone it continues to enlarge both joints and bedding planes until caves are formed. These vary in size but are often spectacular – giant caverns below the ground. These caves become enlarged by solution and by the rush of flood water through them after heavy rainfall. The shape of the underground caves and passages often depends

Fig 15 The formation of a dripstone

Fig 16 Stalactites form on the ceiling, stalagmites form on the ground.

on the pattern of joints or on their depth below the surface. Deeper passages, below the water-table, are generally more rounded and tube-like. Streams in them flow at greater pressure. Passages nearer the surface usually only contain streams when the water-table is very high. They are often shaped like a key-hole in cross-section as running water only wears away the floor of the passage.

Fig. 15 shows how limestone caves can be formed. Water seeps through the limestone joints above the cave, carrying dissolved calcium carbonate with it. As it drips from the ceiling of the cave some of the carbon dioxide escapes and the water is unable to hold all the calcium carbonate. This is left on the ceiling or wall. As the drop of water falls to the floor more calcium carbonate is deposited. These deposits are all known as *dripstone* – the best known of which are the stalactites and stalagmites.

Fig 17 Gordale, near Malham, North Yorkshire

Although much water in limestone passes through rock joints and flows underground there may be some streams on the surface. These may occur where the flow of water is too great for the size of the joints. Some streams use their former surface bed as an overflow channel in times of heavy rainfall.

Flat-topped plateaux and deep gorges are common in limestone areas. As there are few surface streams horizontal limestone strata tend to be worn down evenly by weathering. This helps maintain a flat plateau surface. There are limestone plateaux in the Mendip Hills in Somerset and near Ingleborough in Yorkshire. Steep-sided gorges are also common in limestone country (see Fig. 17). When rivers *do* flow across limestone they often cut vertically into the rock by solution and the normal valley-forming processes described in Chapter 5 do not operate.

(a)

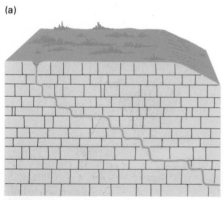

Water seeps slowly into joints and bedding planes

(b)

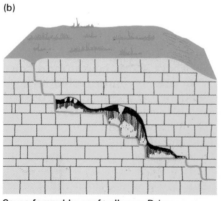

Caves formed by roof collapse. Drips from ceiling

(c)

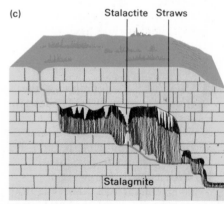

Water continues to drip in cavern. Different types of dripstone formed.

Slopes

Describing slopes

All landforms are made up of slopes. Slopes, however, are difficult to describe. Think about hills near your home. How would you describe them – steep?, gentle?, undulating? What is steep to you might not be so steep to someone else. Geomorphologists have tried to describe and measure slopes accurately. The morphological map exercise in Chapter One (p. 2) showed one way of describing slopes, but it is sometimes useful to make more accurate measurements. Fig. 18 shows a *slope pantometer* in use. This is a simple instrument used for measuring slope angles. It consists of a frame (1 m × 1 m) loosely bolted at each corner. A protractor scale is fitted to one of the uprights and a spirit level 'bubble' attached with spring clips to show when the frame is vertical. When the pantometer is placed on sloping ground the angle of the slope may be read off the scale. The pantometer is moved up the slope, a section at a time, and the angles recorded. These can then be plotted on graph paper to give an accurate profile of the slope. Slope angles are usually measured at right angles to the contours.

Fig 18 Measuring slopes with a slope pantometer

To help in the description of landforms, names have been given to the parts or 'elements' of slopes. The names given to some common slope shapes are shown below.

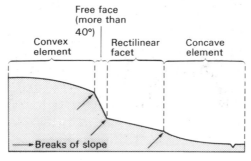

Fig 19 Naming the parts of slopes

12 The following notes (Fig. 20) were taken from a student's field notebook while he was measuring the valley slopes of a small stream – the Wigg Beck – with a slope pantometer. Using these notes construct a profile of the valley side. Use a scale of 1 cm to represent each 2-metre unit of the real slope. This has been started for you in Fig. 21. Label the profile to show breaks of slope, convex and concave elements and rectilinear facets.

Slope Survey of Wigg Beck Valley River

Slope pantometer readings up valley side :

| 40 (degrees) |
| 1 |
| 10 |
| 14 |
| 20 |
| 27 |
| 48 |
| 0 |
| 0 } level ground |
| 0 |
| 3 |
| 57 |
| 39 |
| 9 |
| 1 |
| 11 top of valley side |
| -2 |
| -20 |

Fig 20 A field notebook recording slope measurements

Fig 21 Use this diagram for question 12

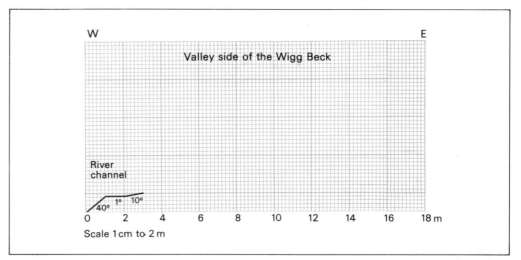

The description and measurement of slopes is very important but geomorphologists today are usually more interested in how slopes change and the processes responsible for their different shapes. The major process at work on hill slopes is the *transportation* of the layer of weathered material. This may be dissolved in the water running off the slope or may move downslope by gravity.

Fig 22 The Vaiont Dam Disaster

DAM DISASTER FLOODS PIAVE VALLEY WHOLE VILLAGE SUBMERGED

MANY CASUALTIES FEARED
Bellino, Italy, Thursday

A vast flood of water from the Vaiont Dam – the third highest concrete dam in the world – swept down the Piave valley just after midnight this morning, pouring into several counties in its path. First reports spoke of a major disaster. Communications into the area were completely wiped out. Police here, about ten miles to the south, first said that the 873 ft high dam had collapsed.

Later reports said that a gigantic landslide down an adjacent mountain had plunged into the reservoir behind the dam pushing thousands of tons of water over the top. Longarone, a county of more than 2000 people close to the dam, was reported by police to be completely submerged.

The civil authorities here and in Venice, about 65 miles south of the dam area, reported that in the past few days thousands of tons of earth and rock on a mountain above the dam had been loosened by torrential rain.

a) extract from *The Times*, 10 October 1963

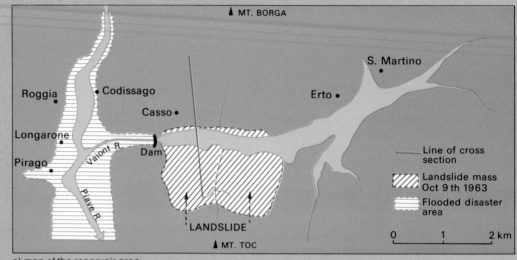

c) map of the reservoir area
d) cross-section of the reservoir

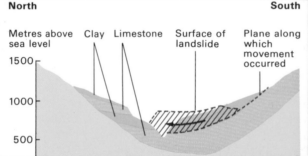

Report of an Engineering Geologist after the Disaster

There is a long history of landslides in this area especially on the north-facing slope of the valley. The Vaiont canyon is cut into a syncline of sedimentary rock. The rocks are composed of alternate layers of limestone, clay and marl (a soft, limey clay). The beds dip steeply towards the floor of the valley. Heavy rainfall occurred in the fortnight before the disaster. The water seeped into the rocks, adding to their weight and lubricating the boundaries between the rock types. In mid-September earth movements of about 1 cm per day were recorded. It was thought, though, that this was the movement of individual blocks of rock and not the whole mountainside moving as a mass.

On about 1st October animals grazing on the northern slopes of Mount Toc sensed danger and moved away. The mayor of Casso ordered the evacuation of the town as a small slide was anticipated. Reservoir engineers began, too late, to lower the water level. The fatal slide occurred at 10.40 p.m. on the 9th October.

The disaster demonstrates the importance of geological conditions in dam building. Under certain conditions rock masses can weaken suddenly and movement rapidly accelerates once begun.

b) the geologist's report

(Report based on Kiersch, G.A. 'Vaiont Reservoir Disaster' *Civil Engineering* v.34, 1964, pp 32–9)

e) the dam before the disaster
f) the disaster scene

Landslides

13 Study the information on the Vaiont dam disaster in Fig. 22 opposite. Locate the area on an atlas map.
   a Describe the geology of the valley.
   b What caused the disaster? Reconstruct the series of events that led to the landslide.

Movements of hillsides are obviously of great importance to geomorphologists. If we can find out more about the way in which such movements take place we can take steps to predict when they will occur, to try to prevent them and reduce the loss of life and damage to property.

14a Place a brick on a board that has been covered with a thin layer of dry soil. Tilt the board and with a protractor measure the angle at which the brick begins to slide. Thoroughly wet the soil and repeat the experiment. Is there a difference between the two angles? Can you explain your result? Imagine the brick and board represent a house built on a slope. What might happen to the house after heavy rainfall? How do builders prevent this?

   b Cut one end from a shoe box. Fill the box with damp sand and tip the box as though it were the back part of a tipper truck. Repeat this several times. Look at the face along which the sand slips. This is called the sheer plane. Is it flat or slightly curved? Draw the typical shape that the sand makes as it tips from the box.

Landslides happen when solid rock or weathered rock fragments on slopes are not properly supported. The landslide at Vaiont occurred because the limestone slipped on the underlying layers of clay. Landslides may also occur because undercutting at the base of a slope may leave too steep a slope angle. Undercutting may be by a river or the sea but is often caused by man. Oversteep motorway cuttings or piles of pit waste may put too much of a strain on a slope. When this happens the slope 'fails', the material slips and settles at a new, shallower angle.

Fig 23 Soil movement on a hillside

The layers or rock fragments produced by weathering are liable to the force of gravity. This movement downslope of weathered material is called *mass-movement*. It is only rarely as disastrous as that in Italy in 1963. Most mass movement is very slow, yet it is the slowest movements that move the greatest amount of material.

Fig 24 Soil creep is helped by frost

Soil creep

15 Look at Fig. 23. Describe what has happened to the features at the surface.

Weathered surface material in the diagram has moved downslope. The slowest type of mass movement is called *soil creep*. Fig. 24 shows how this process is helped by frost. Soil creep is too slow to be observed directly. The evidence for movement in Fig. 23 has developed slowly over many years.

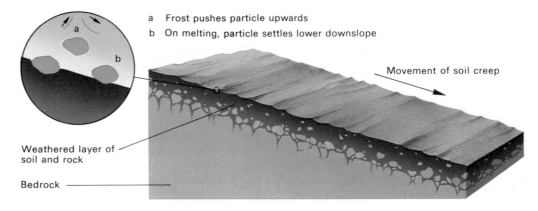

a Frost pushes particle upwards
b On melting, particle settles lower downslope

Movement of soil creep

Weathered layer of soil and rock

Bedrock

Fig 25 A small mudflow

Labels on Fig 25: **Overhanging turf**, **The brown mud here has flowed down the slope**

Slide

Slump

Direction of slide

Direction of slump

Surface of slumped block tilts backwards

Sliding occurs along bedding plane

Slumping along a curved 'sheer plane'

Original surface   After landslide

Fig 26 Sliding and slumping – two types of landslide

Fig 27 Cain's folly, landslides at the coast near Charmouth, Dorset

When soil is wet there is greater likelihood of movement. If the surface layer of soil is soaked, the layer itself may slowly flow downslope as a soggy mass. This is called *solifluction*, which means 'soil flow'. With more water and on steeper slopes the flow is faster – it is then called a *mudflow*. A small mudflow is shown above.

Look at Fig. 26 which illustrates two different types of landslide – a *slide* and a *slump*. Both of these are common where strong rock overlies a weaker rock such as clay. Slides and slumps are usually fairly rapid although they may vary in size. Some slides may be of great size – the largest ever was the immense prehistoric landslide at Saidmerreh in South Iran. A block of limestone 14 km by 5 km by 300 m slipped off the Kabir Kuh mountain sending about 50 billion tons of debris rushing into the valley below!

16 Refer to the photograph of 'Cain's Folly'. Make a tracing of this section of the coast near Charmouth in Dorset.

Add labels to your tracing to show:
a where undercutting by the sea takes place
b where the sea is removing debris from old landslides
c where landslides have recently occurred.
Shade in a different colour the scars left by landsliding.

On cliffs and steep mountains pieces of rock weathered by frost shattering fall freely as *rock fall*. The broken pieces pile up on a slope at the foot of the cliff. This slope is known as a *scree*.

17 Make an annotated sketch of the screes at Ennerdale in the Lake District. Label your drawing to show where frost-shattering occurs.

18 The types of mass movement in this chapter have been classified according to their speed of movement and the amount of water needed for movement to take place. Mark each of the following processes in its correct position on the graph on the right:
(Number 1 has been done for you)
1 landslide
2 soil creep
3 solifluction
4 mudflow
5 slumping
6 rock fall
Write a sentence about each of these types of mass movement.

Review Questions

1 Which of the following words describe Carboniferous Limestone?
permeable
impermeable
porous
pervious
non porous
impervious
2 Explain the difference between the following pairs of words:
clints, grykes
stalactites, stalagmites
swallow hole, resurgence
3 Look at Fig. 30.
What type of rock is shown in the photograph?
What chemical does this rock consist of?
Estimate the thickness of the soil and describe the weathering of the rock near the surface.
Explain the weathering process that is occurring and say why the soil layer is so shallow.

Fig 28 Screes at Ennerdale, Cumbria

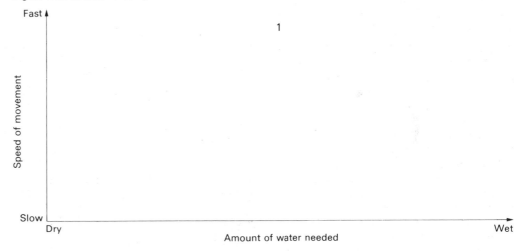

Fig 29 Classification of mass movement (use for question 18)

4 Say how a study of mass movements would be of value to:
– a mining engineer planning the dumping of waste material from a coal mine
– a civil engineer building a road in a mountainous area
– a property developer proposing a new housing estate near clay cliffs.
For each case say what you think the dangers or problems might be for each situation and how you might be able to solve them.
You could do this exercise either as a piece of written work or as group discussion with each group producing a report.

Fig 30 Weathering on the South Downs

5 Rivers

Fig 1 Rills on a hillside

1 On an outline map of the British Isles, mark and name the following rivers: Thames, Severn, Trent, Exe, Tees, Tyne, Tweed, Clyde, Forth, Shannon, Foyle, Dee (N. Wales), Dee (Scotland).

2 Study an Ordnance Survey map of your local area. Identify any towns built at bridging points of rivers and any low lying areas where there are few settlements.

River basins

Study Fig. 1. The marks on the hillside are *rills* – tiny streams where the water is collecting and moving downhill. They join with other rills to form streams which then meet other streams to become a river. So a river has many streams and rills flowing into it. These are known as *tributaries*.

If rain falls anywhere within the red line on Fig. 2 it will end up in the main river. This area is known as a *river basin* or

drainage basin. The line which separates one drainage basin from another is called a *watershed* or *divide*.

3 Trace Fig. 3. Draw in the watersheds of the river basins draining to A, B and C. The watershed of the river flowing to D has been drawn to show you the method. Notice how watersheds follow the hills and ridges around the valleys.

Fig 2 A river, or drainage basin and its watershed

The British landscape owes many of its distinctive features to the work of rivers. Hills and valleys generally exist because running water has cut down into the rocks. Mass movements (Chapter 4) move material down these slopes and the valley's shape may have been modified by glaciation (Chapter 6) but rivers may be seen as a basic influence on our landscape. They are also of great everyday importance to us. They provide water for our homes and for use in industry. The larger rivers are used for carrying goods by barge and many of the smaller rivers are visited by us in our spare time for fishing or boating. In extreme conditions of droughts or floods rivers reach the headlines of our newspapers. Historically towns have grown up where important routes meet to cross a river at a bridging point. On the other hand some low-lying areas have few villages or towns because of the danger of flooding.

Fig 3 Watersheds (use this map for question 3)

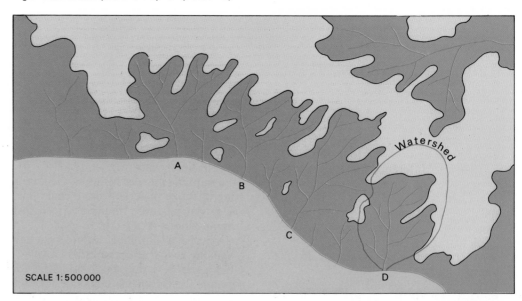

SCALE 1:500 000

4 Make a model of the land shown in Fig. 3 using wet sand (or your school could make a more permanent model out of plaster). Make 'rain' over the model using a watering can or hosepipe with a 'rose' attached. Study the route taken by the water as it flows into the rivers. Did you draw in the watersheds correctly when answering question 3?

5 Drainage basins may be very large or very small. Notice that each of the rivers shown in Fig. 3 has tributary streams which have their own drainage basins. Add some of these smaller basins to your copy of the map

Stream order

When a stream flows away from its *source* – whether this is a spring, a glacier or a lake – it is known as a first order stream. When this is joined by another stream the result is a second order stream and it remains so until it meets another second order stream and becomes third order. This system of numbering allows us to refer to a river or river basin in precise terms.

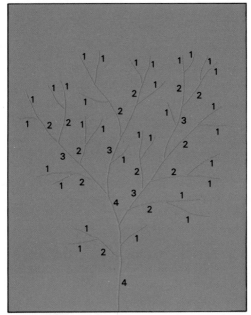

Fig 4 Stream orders

6 In Fig. 3, what order of river enters the sea at A? At B? How many first order streams are there in each of these basins?

7 Trace the streams shown on an Ordnance Survey map of your local area or an area you are studying. Write in the stream orders. What is the largest order of stream shown in the area? Draw in the watersheds of two second order drainage basins and shade the basins they enclose.

Drainage patterns

The arrangement of streams within a drainage basin reflects local conditions of climate, geology and relief. Sometimes streams are near together so that all areas of the basin are close to a stream. This particularly happens where there is heavy rainfall or impermeable rocks. Such an area is said to have a high *drainage density*. Drainage density is calculated by adding up the length of all the streams in a river basin and dividing by its area. For example, in Fig. 3.

$$\text{Drainage Density of River Basin A} = \frac{54\,\text{km}}{46\,\text{km}^2} = 1.17\,\text{km/km}^2$$

The shape as well as the density of streams varies from basin to basin. A *dendritic* drainage pattern resembles the shape of the trunk, branches and twigs of a tree and its name comes from the Greek word, *dendron*, for a tree. This pattern is found where there are few variations in rock type. Streams flow into each other almost at random rather than picking out lines of weakness. Most clay lowlands in Britain show this pattern on a local scale. You can see it, for example, in the south-

Fig 5 Trellised drainage

ern part of the River Mole's drainage basin. (Fig. 8).

8 Study the Ordnance Survey map of your local area or of the nearest area of countryside and see if you can identify a river basin with a dendritic drainage pattern.

9 Trace the pattern of rivers in the Lake District from an atlas map and shade the high land.

The Lake District has a *radial* drainage pattern. The rivers follow a shape similar to radii drawn from the centre to the outside of a circle. This pattern is usually found where rivers drain away from a dome shaped area of hills.

Trellised drainage occurs where there are alternating bands of hard and soft rocks. The hard rocks often give rise to escarpments with streams forming valleys along the soft rocks between. Only occasionally do streams break through the escarpments. This pattern of right angled bends and more or less parallel lines looks rather like a garden trellis.

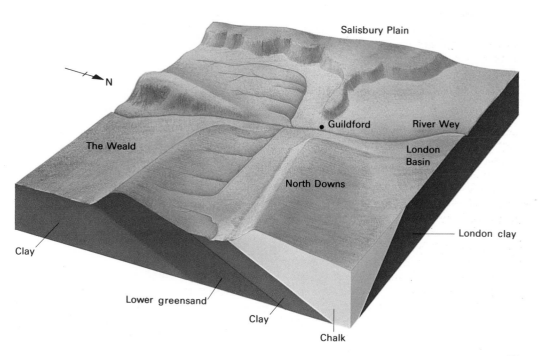

The water cycle

We have seen (p. 4) that the water cycle is a *system* involving water in a liquid form (oceans, lakes, rivers, clouds and rain), solid form (snow flakes, glaciers) and as a gas (water vapour). Sometimes the water changes rapidly from one state to another; it may evaporate from the sea to become a gas which in turn condenses to form clouds. Rain may then fall on the land to form the major *input* of water to the drainage basin. Water leaves the basin (*output*) by evaporation or by rivers flowing into the sea.

Sometimes instead of moving rapidly through the water cycle, water may remain in one part of it for a very long time. For example, it may be *stored* as glacier ice or be trapped deep in the ground for thousands of years. If we look at a river basin in more detail we shall see that there are many routes from the input, rain, to the main output, the river. Some routes are fast and others slow. The factors involved all influence the *regime* of the river – variations in its size from day to day and from season to season.

Rainfall

This may be concentrated in one season of the year as in the monsoon of India or the winter of Mediterranean lands, giving rivers which vary greatly in size from one season to another. In Britain we do not have a seasonal drought but we have variations from day to day which cause changes in the size of our rivers. Heavy thunderstorms in particular are likely to cause flooding.

Interception

Where there is dense vegetation some of the rainfall never reaches the ground. Water is trapped on the leaves and evaporates. Water also reaches the ground over a longer period of time as it may still be dripping from the leaves hours after the rain has stopped. In both these ways, dense vegetation decreases the risk of flooding, and we should bear this in mind when we replace forests with grass, arable land, roads or houses.

10 To investigate interception by vegetation.

a Construct a number of rain gauges to measure the rainfall. These may be very simple, such as a plastic funnel in a milk bottle – but they should all be of the same design.
b Record rainfall in a gauge sited in an open area. Also put a number of gauges beneath trees and record rainfall in these.
c Compare the rainfall recorded by the gauge at the 'open' site with the average from the gauges beneath the trees. Calculate what percentage of rainfall is lost by interception in the wooded area.

Surface runoff and infiltration

When water hits the soil it may sink in (*infiltrate*) or run over the surface (*runoff*). If the rainfall is very heavy the ground may not be able to absorb it quickly enough, resulting in a large proportion of runoff. This water will reach the river quickly causing a rapid increase in its size and a high risk of flooding. The amounts of infiltration and runoff are also influenced by the nature of the soil.

11 To investigate infiltration rates.
a Open a tin can at both ends, being careful not to make jagged edges.
b Draw a line round the inside and outside of the can 2 cm from the bottom. Then draw a scale on the inside of the can as shown in Fig. 6.

Fig 6 Measuring infiltration

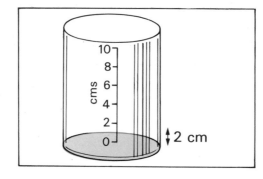

c Select a number of contrasting sites e.g.
raised flower bed; compressed soil on playing field; sandy soil; clay soil; wet soil; dry soil;

At each site, drive the tin can into the ground (using a hammer and a piece of wood if necessary) up to the 2 cm mark.
d Pour in water up to the mark 10 cm above the base line. Time how long it takes each cm of water to drain into the ground. This may vary from seconds to hours!
e Draw up your results as a series of graphs similar to Fig. 7.

Fig 7 Recording the results

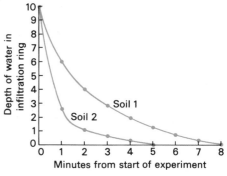

f Explain the differences between your graphs. In a storm, which site would produce (i) most (ii) least run off? Why is the rate of infiltration often faster for the first cm of water than for the last one?

12 Which is likely to produce large quantities of runoff:
steeply or gently sloping ground?
soil which is dry or wet when the rain starts?
soil which is loose or compact?
soil which is permeable or impermeable?
In each case, can you say why?

Movement of water through soil and rock

Many streams start as springs fed by water from underground. Water also enters rivers directly from the soil along the river banks. When water moves quickly through the soil and into a river it helps to produce flood conditions. But if it is absorbed by porous rocks it may take months to get to the river and the effect of the storm will hardly be noticed in the size of the river's flow.

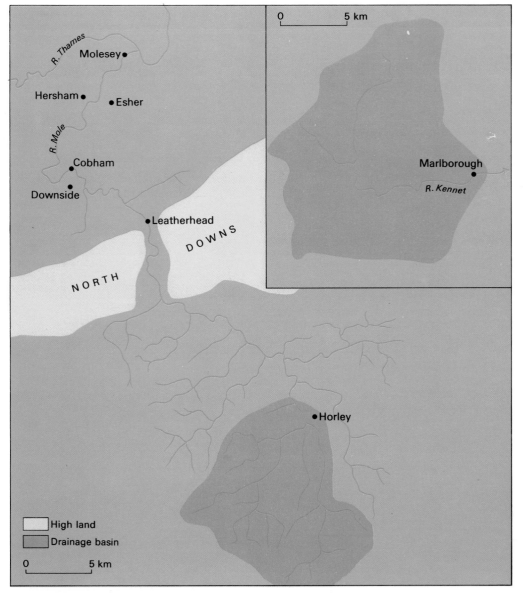

Fig 8 The drainage basins of the River Kennet and
the River Mole

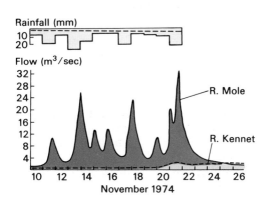

Fig 9 Hydrographs of the River Kennet and the
River Mole

Evaporation and transpiration

Surface runoff and moisture in the soil may never reach the river. Instead it may leave the system directly into the air, either from the ground (*evaporation*) or through the leaves of plants (*transpiration*). This is especially likely in hot, dry and windy conditions.

River flow

This is affected by all the features of the drainage basin we have been considering. If water moves quickly through the system this is likely to result in a river which experiences extremes of floods and droughts. If water moves through the system more slowly the river flow will be evened out.

Floods

Fig. 8 shows the drainage basins of the River Kennet and River Mole, both tributaries of the Thames. The Mole up-stream of Horley flows over impermeable clay whereas the Kennet flows across porous chalk. Because water sinks into the chalk drainage basin, this has a much lower drainage density than the clay area. The nature of the rock also influences the *flow* of the rivers. The Mole responds rapidly to any rain which falls. Rainfall is immediately followed by an increase in the river's size and the flow soon decreases again. This is because water reaches the river rapidly as surface runoff or through the soil and is quickly moved out of the system. In the Kennet basin there is little response by the river to individual rain storms because water sinks into the rock rather than reaching the river quickly as surface runoff. Only towards the end of the period shown is there a slow increase in the river's size. Overall the flow is far more steady than that of the Mole and a river like this has fewer floods or droughts. It is a reliable source of water for homes and industries. The Mole is not nearly so well behaved.

THE MOLE VALLEY FLOODS

Heavy rainfall on 14 and the 15 September caused serious flooding throughout the Mole valley. It was particularly severe in low lying areas near the confluence with the Thames. This newspaper article describes the sequence of events.

FRIDAY – It starts on the Friday before the floods, four hundred miles away over the holiday beaches of the Scilly Isles. A depression builds up, and moves first towards Biscay, then in over France.

SATURDAY – The depression extends to Southern England, and is expected to pass northwards. But instead it stays still, blocked by another weather build-up. It brings warm air and thunder, and the rains begin to fall.

SUNDAY – It is still raining, with over double the amount for the time of year, Roads in Surrey, and especially in the Mole valley, become impassable.

AT 3.a.m. the main A3 road is under six feet of water. By lunchtime the Mole has changed the Cobham landscape with floods half-a-mile wide. Houses are evacuated near Downside Bridge. Soon it will disappear, swept away by a wall of water.

Leatherhead and the other towns along the Mole are swamped. In Esher, Molesey and Hersham, people sit in warm living rooms watching television pictures of floods further up the Mole valley. As yet, they know nothing of what is happening even in Cobham. As they sleep during Sunday night, the rain slackens. But the damage has been done, and from now on the waters of the Mole dictate terms.

MONDAY – Now, the combined flood waters from the length of the Mole are starting to reach the final stage of their journey to the Thames. Around 5.45 a.m. people in Riverside Drive, Esher, next to the Mole wake up to hear water lapping downstairs. It is only inches deep, yet even as they struggle to pile furniture upstairs, the water deepens to two feet in an hour. The writing is on the wall for Molesey but even at lunchtime most residents still don't realise what's going on. There has been no official warning, more than a day *after* the Mole flooded parts of Cobham. By 4 p.m. West Molesey is virtually a lake. By evening the waters start to take a hold on East Molesey, and rise steadily through the night.

TUESDAY – The worst flooding so far but the floods begin to recede in Cobham and Hersham. By FRIDAY the Mole is back in its course.

Fig 10 The Mole floods, September 1968

13 How many days passed between the start of the rain and the worst flooding at Molesey? Why were the floods still deepening a day after the rain stopped?

14 Did the inhabitants of riverside areas in Esher, Molesey and Hersham really need an official warning, or could they have expected flooding when they saw television pictures of conditions in Leatherhead and Cobham?

Before the floods large sums of money had been spent on flood control along the Mole. Sluices were built – gates across the river which can be shut to hold back the water or opened to let it through. Flood banks had been raised and changes made to the river channel. Why were these measures not successful? One reason was that the sluices were not properly operated, as the storms caught the area by surprise. Better weather forecasts might help prevent flooding. Also, the towns along the Mole valley had been growing.

Large areas of buildings and roads had been constructed. In such areas water, instead of being absorbed by soil, now runs rapidly through the drains into the rivers, increasing the chance of flooding. Finally, without spending millions of pounds it is not possible to protect low lying areas against really freakish storms. Since 1968 further improvements have been carried out but this still does not guarantee protection against flooding in the very worst conditions.

15 Put yourself in the position of the mayor of New Daigtown. The town is subject to flooding and you have obtained estimates for measures that might be taken to control this. These measures include:

a Constructing dams on the headwaters of the river so that floods can be held back by sluice gates.

b planting trees in the headwater area.

c raising the height of the river banks.

d straightening the course of the river so that water flows away more quickly.

The estimates are as follows:

For a flood that might happen every	The damage done each time would be	The cost of flood control would be
20 years	£10 million	£8 million
30	£35	£27
50	£55	£50
100	£100	£124
1000	£800	£1350

You should bear in mind;

that your council would have to raise this money from the present inhabitants of the town

that the flood which could happen every 100 years might happen tomorrow – or in a hundred years' time

that the 100 year and 1000 year floods would result in loss of life if they were allowed to happen

that when campaigning for the last election you promised to keep council spending to a minimum

that the flood control measures would have an estimated 'life span' of 100 years. After that, major repairs and rebuilding would be needed.

a Write your speech to the town council, saying which frequency of flood you think the council should guard against, and why you have come to that decision.

b Why do you think the consultants suggested planting trees as a way of aiding flood control?

c If the council refused to accept any of the consultant's schemes,

i) would you encourage people to move away from any – or all – of the flood zones shown on the map?

ii) what sort of new land uses would you encourage in the zone which is likely to be flooded every twenty years – parks? industries? houses? car parks? or something else?

Recording river discharge

We have seen that the flow of a river is of great importance, particularly in extremes of drought or flood. If we want to be able to forecast these or plan to prevent them we first need to keep records of river *discharge*, that is the amount of water flowing past a point on the river bank in a given length of time. *Gauging stations* which keep these records have a specially built section of channel called a *weir* through which the water passes. Where there are no gauging stations, other methods have to be used and it is quite possible to carry out such measurements with the simplest of equipment.

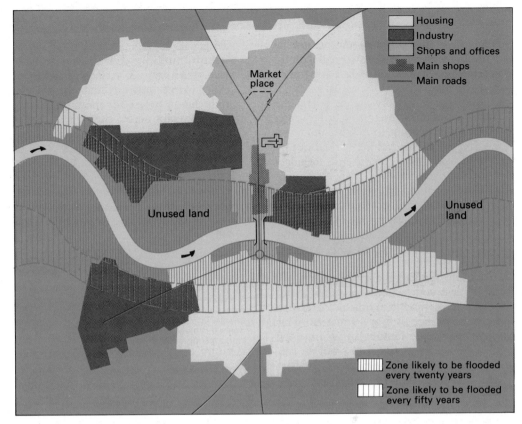

Fig 11 Map of New Daigtown

Fig 12 A river gauging station to record discharge

51

16 A practical exercise to investigate the discharge of a stream. Note: for safety's sake this practical should not be carried out in streams over half a metre deep, or when water is fast flowing or very cold.

a Discharge is usually measured by the number of cubic metres of water going past a given point in one second. You will see that this can be worked out by multiplying the river's *cross sectional area* by its *speed*.

b Select a practical place to measure stream discharge – somewhere where the water is neither too deep nor too fast and where there are several metres of stream flowing at a roughly constant speed.

c Measure the average speed of flow:
i) Unless your school has a flow meter the best method is to time a float over a given distance of 5 or 10 metres. This method, made famous by Winnie the Pooh, is not as accurate, but is certainly good fun! The best sort of float to use is something easily seen and which floats well down in the water – an orange, a fishing float or suitable piece of wood. If too much of the float stands up above the surface the wind may blow it upstream, suggesting that the river is flowing in reverse!
ii) Time the float over the measured distance, putting it in at several points across the river. The speed of the river is usually fastest in the middle and slowest at the sides, so a number of readings will allow you to work out the *average* surface flow of the river. Express your answer in metres per second.

iii) The surface speed is likely to be faster than the flow near the river bed. To get an average speed for the whole stream, multiply your answer for surface speed by 0.8. A large number of experiments suggest that this is the right figure to use for most rivers.

d Stretch a tape across the stream. Every half metre from one bank, measure the depth of the stream, using a ruler or measuring stick (Fig. 15). Record these results carefully.

e Draw a cross-section of the stream onto graph paper. Use the same scale for the width and depth of the stream. Work out the area represented by each square of the graph paper. Count the number of squares making up the cross-section of the stream. Calculate the area this represents in reality, expressing your answer in square metres (Fig. 13).

f Multiply the cross-sectional area (in square metres) by the stream's speed (in metres per second) to give you the discharge (in cubic metres per second).

This practical may be extended to investigate how fast the discharge of a stream increases downstream from the source, or

Fig 15 Measuring the discharge of a small stream. Here the cross-section of the stream is being measured.

Fig 14 Plotting discharge on a hydrograph

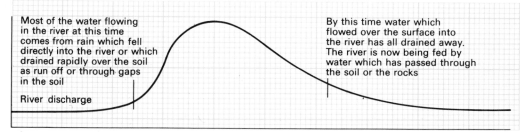

Most of the water flowing in the river at this time comes from rain which fell directly into the river or which drained rapidly over the soil as run off or through gaps in the soil

River discharge

By this time water which flowed over the surface into the river has all drained away. The river is now being fed by water which has passed through the soil or the rocks

Time from start of storm

to see how discharge varies from day to day (which may then be compared with a record of the rainfall).

If discharge is measured over a period of time, the results may be plotted on a *hydrograph*. Typically, after a rainstorm the river rises fairly rapidly as runoff and water moving quickly through the soil reach the channel. The flow decreases less quickly as, though much of the storm water has been carried away, water is still percolating into the river from the soil and from porous rocks.

Fig 13 Measuring discharge

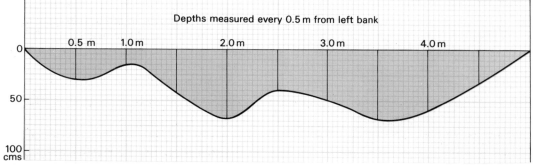

Depths measured every 0.5 m from left bank

0.5 m   1.0 m   2.0 m   3.0 m   4.0 m

0

50

100 cms

Each small square $= 5\,cm \times 5\,cm = 25\,cm^2$
Number of squares $= 810$
∴ Area of stream $= 810 \times 25\,cm^2 = 20\,250\,cm^2$
$= 2.025\,m^2$

Average speed of stream $= 0.20\,m/sec$
∴ Discharge $= 2.025\,m^2 \times 0.20\,m/sec$
$= 0.405\,m^3/sec$

17 Plot a hydrograph in the form of a bar graph, using information from Fig. 16.

We have been studying the reasons for variations in river discharge and particularly for flooding. In the next section we shall see that discharge is a very important factor in influencing the nature of a river's channel and the valley in which it flows.

River valleys

A river's energy

In our everyday life we use our energy for lifting things, or for walking and running. The energy comes from the food we eat. An athlete who uses a great deal of energy often has to eat more so that he has this energy. A river is in some ways similar. It uses energy to erode its bed and banks and to transport its load of pebbles and fine material. If it is to carry out a great deal of this work it needs a lot of energy. A river's energy depends on the *slope* of the land and the *amount of water* flowing along it, so we can say that the steeper the gradient and the larger the river, the more energy it has.

What is this energy used for? 95 per cent is used simply to keep the river flowing. As water moves over the bed and banks, these tend to slow it down, a process known as *friction*. In a small channel or a channel with very rough bed and banks, friction slows the river even more. A river at low water has less energy than the same river at high water simply because it is smaller. In addition the low water river is having to use up more of its energy in overcoming friction. As a result, the high water river flows much faster, erodes its channel and transports more material downstream.

You might expect a river in its upper course, where the gradient is generally steep, to flow faster than a river further downstream but this is not the case. The upland river has a small, rough channel and uses up more of its energy in overcoming friction whereas less of the water in a lowland river is being slowed down by the bed and banks. The result is that the average speed of a mountain torrent is very

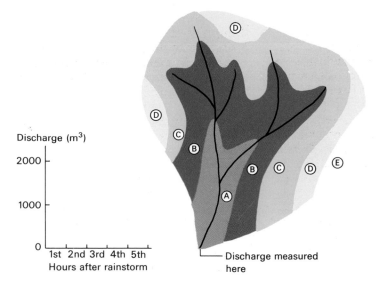

Discharge (m³)

2000

1000

0

1st 2nd 3rd 4th 5th
Hours after rainstorm

Discharge measured here

Fig 16 River discharge

A steady 300 m³/hr. are fed into the stream from porous rocks in the area. In addition to this, after a storm;
500 m³ arrives at the gauging station from zone A in the first hour after the storm.
1500 m³ arrives at the gauging station from zone B in the second hour after the storm
1000 m³ arrives at the gauging station from zone C in the third hour after the storm.
500 m³ arrives at the gauging station from zone D in the fourth hour after the storm.
200 m³ arrives at the gauging station from zone E in the fifth hour after the storm.

similar to that of a large river flowing over a lowland. It is true that in places such as waterfalls and rapids the upland stream may have sections of very fast flow but this is balanced out by eddies where water may be moving downstream very slowly, so the average speed is not as fast as one might suspect.

Fig 17 The speed of water varies within a river

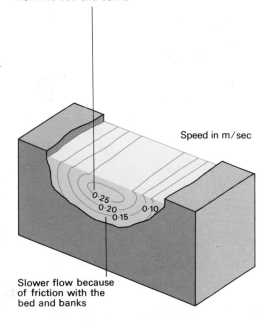

The fastest part of the river is well away from the bed and banks

Speed in m/sec

0.25
0.20
0.15
0.10

Slower flow because of friction with the bed and banks

The 5 per cent or so of energy left over after a river has overcome friction may be used to wear away or *erode* material from the edge of the channel and to *transport* (carry) it downstream. If the river is transporting material and there is a decrease in its energy, the material may be *deposited* (dropped to the river bed).

There is then a balance between the energy a stream has and the work it carries out. The shape of the channel is a result of the balance that has been achieved. The delicacy of this balance is sometimes forgotten and can easily be upset by man. When a dam is built across a river, deposition occurs in the lake it holds back. The lake is gradually filled in. The river downstream of the dam is starved of sediment and it may erode its channel vigorously.

18 A fast flowing river is likely to have more energy for erosion than a slow one. Bearing this in mind, is erosion or deposition more likely when
a a river flows from a steep to a gentle gradient?
b heavy rainfall causes the river to increase in size?
c a river speeds up because the channel is smoother and larger?
d a lot of material falls in to the river when a section of bank collapses?

Erosion

Processes of Erosion

Moving water erodes best when it is carrying particles which it throws against the bed and bank. These in turn dislodge more particles, a process known as *abrasion* or *corrosion*.

Fig 18 Potholes in a river bed. The hammer is about 30cm long

19 Study Fig. 18. Describe these *pot holes* in the river bed, mentioning their shape and estimating their depth and width.

Pot holes are formed where a hollow in the stream bed is deepened by a stone whirled round and round by the force of the water. By itself, water can only erode soft material unless it is travelling very fast. This process is known as *hydraulic action*.

20 Use a dictionary to find out the meaning of the word 'hydraulic'. What other words start with 'hydra' or 'hydro'?

A river may also dissolve its bed and banks. Rivers flowing over chalk or limestone carry out much of their erosion in this way (*solution*). As the load of a river is carried downstream, particles knock against each other and also hit the bed and banks. In doing so, they erode each other, becoming smaller and more rounded. In this process, known as *attrition*, material is gradually worn down to form fine rock particles called *silt*.

Results of Erosion

Erosion may be mainly *vertical* or *lateral* (horizontal). In upland areas most erosion takes place vertically on the bed of the river as shown by the occurrence of pot holes, or of *waterfalls*. Waterfalls or rapids may form where a stream plunges into a deep valley formed by a glacier in the Ice Age, where it crosses a fault line or where it flows from hard to soft rock.

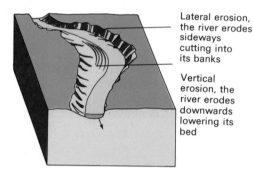

Lateral erosion, the river erodes sideways cutting into its banks

Vertical erosion, the river erodes downwards lowering its bed

Fig 19 Vertical and lateral erosion in a river

Fig 20 High Force, Teesdale

21 Find out the names and heights of some of the world's largest waterfalls.

22 How to investigate the formation of a waterfall. Make a slope of damp sand in a plastic tray as shown in Fig. 21. This layer of plasticene represents a layer of hard rock. Make a stream channel down the slope and pour water gently into the head of the channel.

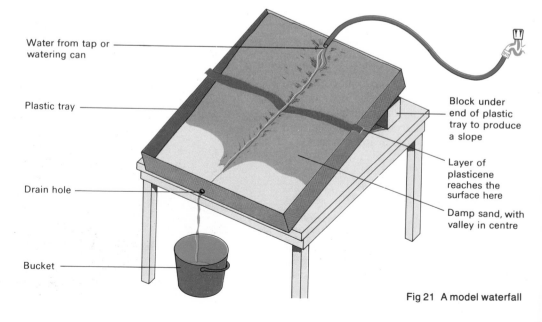

Water from tap or watering can

Plastic tray

Drain hole

Bucket

Block under end of plastic tray to produce a slope

Layer of plasticene reaches the surface here

Damp sand, with valley in centre

Fig 21 A model waterfall

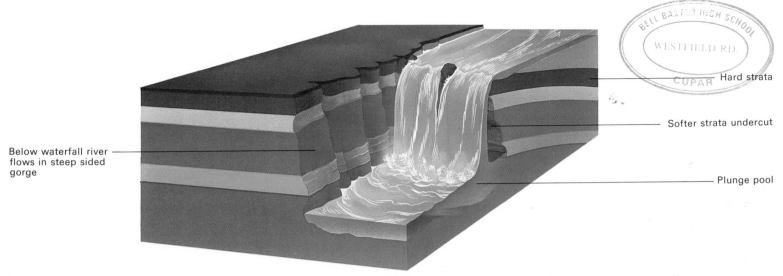

Below waterfall river flows in steep sided gorge

Hard strata

Softer strata undercut

Plunge pool

Fig 22 The features of a waterfall

Observe the formation of a waterfall and sketch the features associated with it, comparing them with Fig. 22.

Erosion in uplands forms valleys which are narrow and steep sided. If downward erosion is very rapid or the valley sides are very hard, a *gorge* with almost vertical sides may result. More often the sides slope down to the river like a letter V, which is why many upland valleys are described as *V-shaped*.

The effect of vertical erosion is also seen in upland areas where a river is swinging from side to side. It cuts down into the land, producing *interlocking spurs*.

23 Fig. 23 shows an upland stream which has cut into the surrounding hills.
 a calculate the gradient of the stream
 b draw a cross-section of the valley from A to B. Would you describe this as a gorge or a V shaped valley?
 c do you think you would be able to see straight up the valley (as in Fig. 24) or would the view be blocked by interlocking spurs as in Fig. 25?

In lowland areas where the river is nearer to sea level it cannot erode downwards very much. Vertical erosion is slower and the effects of lateral erosion are more easily seen. Most lateral erosion takes place where a river is swinging from side to side or *meandering*.

B

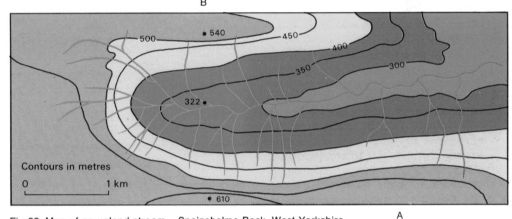

Contours in metres

0   1 km

Fig 23 Map of an upland stream – Snaizeholme Beck, West Yorkshire

A

Fig 24 A V-shaped valley

Fig 25 Interlocking spurs in a river valley

24 Draw a sketch map of the river shown in Fig. 26. Your map should show the shape of the river as it would appear if you were looking straight down at it from above. Mark the river cliff and shade the area of recently deposited material which has no grass growing on it. What evidence is there on the photograph that part of the river cliff has collapsed into the river fairly recently?

25 Trace Fig. 27, which shows where the fastest flowing current is found along a meandering river. Why do you think the current sometimes hits the banks of the river rather than staying in the middle of the channel? Mark on your tracing the places where you would expect to find erosion and deposition.

26 To investigate the behaviour of a meandering river, set up a miniature stream as in Fig. 21 but make the sand slope more gently, leave out the plasticene layer and give the stream a meandering course. Observe the movement of sand particles in the stream channel. Are there particular places where erosion or deposition occurs? Are river cliffs formed as in Fig. 26? If not, experiment with the shape of the meanders, the gradient of the stream and the amount of water you are pouring down the channel.

27 The extent to which a river meanders mostly depends upon its size. The larger the river, the bigger its meanders. The extent to which a river meanders can be described by a simple *sinuosity ratio*:

$$\frac{\text{River Length AB}}{\text{Straight line distance AB}} \times 100$$

With this measure, a straight river has a sinuosity ratio of 100 and the larger the meanders are, the bigger this figure becomes.

a work out the sinuosity ratio between A and B on Fig. 27.

b use an Ordnance Survey map or series of maps to follow a river from its source to the sea. It is true that the meanders get bigger as you go downstream?

Fig 26 Lateral erosion in a meandering river

Fig 27 Where the current flows fastest in a meandering river

Fig 28 Interlocking spurs are eventually removed by river meanders

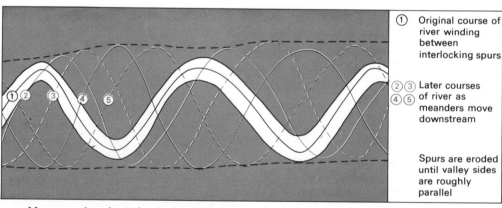

① Original course of river winding between interlocking spurs

②③ ④⑤ Later courses of river as meanders move downstream

Spurs are eroded until valley sides are roughly parallel

Measure the sinuosity ratio near the source, near the middle of the river and near the mouth. Does the sinuosity ratio gradually increase?

The formation of river cliffs by erosion on the outer, downstream side of a meander (Fig. 27) shows that the position of a meander is gradually changing. Over very long periods of time the meanders will bite into the interlocking spurs further and

further until they erode them altogether (Fig. 28). You can demonstrate in a few seconds what a river achieves in a thousand years by lying a length of rope on the ground and flicking one end from side to side. A pattern of waves passes along the rope. Think of these as the meanders changing their course and moving downstream. The area covered by the waves which pass along the rope is similar to the *flood plain* of a river, the wide flat valley floor cut by the meanders as they move down valley. The spurs are cut back to form more or less parallel *bluffs* and the river is no longer flowing in a narrow V-shaped valley.

Fig 29 A flood plain. Notice the bluffs

Fig 31 The formation of ox-bow lakes

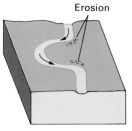

(a) River is meandering. Erosion on the outside of bends leads to the formation of ....

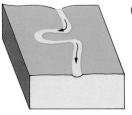

(b) a 'swan's neck' meander.

(c) In time of flood, water takes a 'short cut' across the neck of the meander. If this becomes the main channel ....

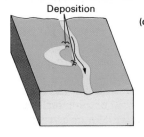

(d) the older channel may be abandoned becoming an ox-bow lake when deposition occurs alongside the new channel.

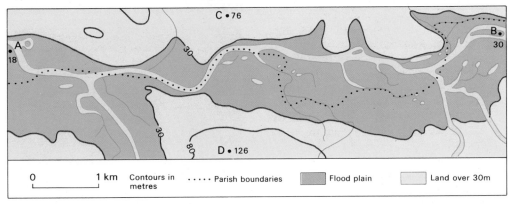

Fig 30 The valley of the River Lune, Lancashire

A close examination of the flood plains of many rivers will produce evidence for the changing patterns of meandering over the years. Fig. 30 shows the valley of the River Lune in Lancashire. Parish boundaries, drawn up about a thousand years ago, would have followed the course of the river. In some places they still do, but elsewhere the river has changed its course. Also notice the streams and lakes on the flood plain. Many of these are the remains of former meanders, now bypassed by the river. Their formation is explained on the left.

In some valleys, *terraces* – abandoned sections of flood plain slightly above river level – are another form of evidence that meanders have changed their course. The river has cut down slightly but has moved a great deal horizontally. Its former routes across the flood plain are shown by the shape of the terraces.

28 a calculate the gradient of the River Lune from A to B.
   b draw a cross-section of the valley from C to D. Mark the present position of the river and label the flood plain and the bluffs.
   c would you be able to see straight up the valley or would the view be blocked by interlocking spurs?
   d calculate the sinuosity ratio of the River Lune between A and B.

Fig 32 River terraces, Swaledale, near Keld, North Yorkshire

Higher terrace

Terrace

57

Transport

Rivers transport their load in several ways which may easily be demonstrated using a piece of guttering.

29 To investigate the movement of material in a miniature river.

a allow water to flow along the gutter. It may appear to be carrying no load but this is not the case. Minerals are dissolved in the water, particularly in areas of chalk or limestone. This is known as *solution load*. Water containing large amounts of chalk or limestone in solution is known as *hard* water. Soap does not lather as easily in hard water as it does in soft water. The 'fur' that sometimes forms in kettles is another sign that the water is hard (see p. 35).

b Drop fine particles of dried and ground up clay soil into your 'stream'. These will not sink straight to the bottom because the flow of water is *turbulent* (Fig. 34). Material carried along in this way is known as *suspension load*. Some rivers are given a distinct colouring by their suspension load – for example the black of the River Negro (a tributary of the Amazon) or the pale colour of the White Nile.

c You may observe some of the larger particles hitting the bottom and then bouncing back up into the stream. Transport in this way, as a series of hops, is known as *saltation*.

d Drop sand particles into the stream. These will be rolled along the bottom as '*traction load*' or '*bed load*'.

Similar experiments could be carried out in a real stream and could then be taken a step further:

30 To investigate the influence of the speed of a stream on the size of the load that may be transported.

a Measure the speed of a stream at several points, to get a range from fast to slow. (Note: remember to take care if investigating a fast flowing section of water).

b Make up several sets of particles, from fine clay through sand and small pebbles to large pebbles.

c Starting with the first material, at each point see which material is moved by the stream and which is not.

d Work out the speed of flow needed to move each type of material.

Rivers erode material from their bed or banks and are supplied with soil or boulders by mass movements on the valley sides. This debris is carried away downstream provided the river has enough energy. But for much of the year little material is eroded or transported because the speed and discharge of the river are too small. Occasionally though, when the river completely fills its channel, it can carry unbelievable quantities of material. Fig. 35 shows a stream descending to Loch Maree in North West Scotland. The large quartzite boulders littering its bed were brought down from the mountains in March 1968 when 16.5 cm of rain fell in 24 hours. Since then they have hardly been moved.

Fig 33 A model of a river transporting its load

Fig 35 Coarse bed load

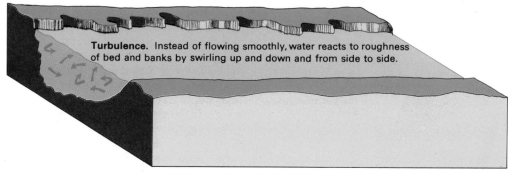
Turbulence. Instead of flowing smoothly, water reacts to roughness of bed and banks by swirling up and down and from side to side.

Fig 34 Turbulent flow in a river

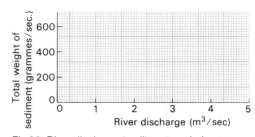
Fig 36 River discharge/sediment carried axes

31 Draw the axes of a graph similar to that above and plot the following figures:

River discharge (cubic metres per second)	Sediment carried (grammes per second)
0	0
1	10
2	60
3	150
4	300
5	580

Would you say that a small increase in discharge makes a small or a big difference to the amount of material a river carries?

Deposition

We have seen how material may be dropped where a river loses the energy to transport all of its load. We shall look at four examples of how this may occur.

Point bars

Refer back to Fig. 26. When a river flows around a meander it erodes material on the outside of the bend but deposits some of its load in the slower moving water on the inside of the bend, forming a *point bar*.

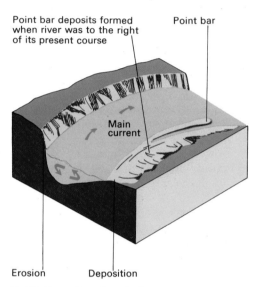

Point bar deposits formed when river was to the right of its present course

Point bar

Main current

Erosion   Deposition

Fig 37 Formation of a point bar

Material on flood plains

If a river bursts its banks it may spread out over a wide flood plain. When this happens there are large areas of silt-laden water moving very slowly over the gently sloping land. Coarse material will tend to be deposited soon after leaving the river channel and this may build up to form *levées* along the banks of the river (Fig. 38). Man has often added to the levées in an effort to build high banks which will prevent flooding.

Further from the river the fine suspended load will also sink to the bottom and be left as a thin deposit of mud. Over the centuries most flood plains have built up considerable thicknesses of silt (*alluvium*) in this way. These areas often form fertile land for agriculture, though it may be used for pasture rather than being ploughed up if it is very damp or frequently flooded.

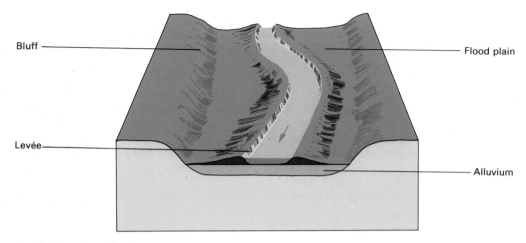

Bluff

Levée

Flood plain

Alluvium

Fig 38 Formation of levées

Deltas

32 To investigate the growth of a delta. Arrange a plastic tray containing sand as shown below. Observe the mouth of the river as you allow water to flow down the channel. Sand is deposited in the water that represents a lake or the sea. Why do you think this happens? The new area of land formed in this way is called a *delta*. Keep a record of its growth, noting changes in its shape and in the pattern of the stream channel.

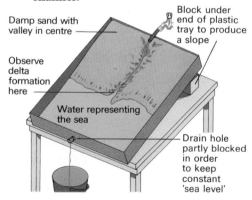

Damp sand with valley in centre

Observe delta formation here

Water representing the sea

Block under end of plastic tray to produce a slope

Drain hole partly blocked in order to keep constant 'sea level'

Fig 39 Modelling a delta

33 Use your atlas to make sketch maps of the Mississippi delta, Nile delta and one other major delta. Note the difference in shape between the Mississippi and Nile deltas. The Mississippi delta is called a '*bird's foot*' delta because of its shape. The Nile delta with its smoother coastline is called an *arcuate*

delta. Note how the river entering a delta splits up into a number of *distributaries*. This often happens because material is deposited in the actual channel of the river, forcing it to split. There are no large deltas around the British coast. Instead, much of the deposition has taken place in estuaries. In the Thames estuary, for example, large areas of mud flats are exposed at low tide and these consist of material deposited by the river.

Alluvial fans

Where a stream flowing steeply through mountains and carrying a big load reaches a lowland it is likely that it will drop much of its load. Over the years this material can build up in a fan shape, rather like a steeply sloping delta at the edge of the lowlands rather than at the edge of the sea.

Fig 40 An alluvial fan

Long term changes

Rejuvenation

An American geomorphologist, W. M. Davis, suggested in 1899 that, soon after a river starts to flow, it develops the steep gradient and narrow V-shaped valley shown in Fig. 23. Comparing rivers with human beings he called this stage 'youth'. Later in the stage of 'maturity' the river develops features similar to those in Fig. 30. Eventually in the 'old age' stage a river wears away the hills almost completely and ends up flowing over an almost flat plain.

Davis realised that when mountains are formed or when the sea level drops these stages of river development would be interrupted and the river would once again start eroding vigorously and taking on some 'youthful' features. This process of making young again is called *rejuvenation*.

Study Fig. 41. The river is meandering widely from side to side but not across a flood plain. Instead, the meanders have cut into the valley – that is they have become *incised*. If sea level falls the river cuts down to the new sea level. This is a very long process but eventually the remains of the old flood plain are left high above the river as *rejuvenation terraces* (Fig. 41). Many British rivers show evidence of rejuvenation because there was a fall in sea level during the Ice Age. At that time more of the earth's water was found in glaciers and ice caps and less of it in the sea.

River capture

Over very long time spans, some rivers lengthen their course by a process known as *headward erosion*. They erode back into the slope above the spring as well as lowering the rest of their valley. This is particularly likely to happen if a river is running along a weak band of rock and such rivers may erode headward into another drainage basin and 'capture' one of the rivers there.

Fig 43 How river capture takes place

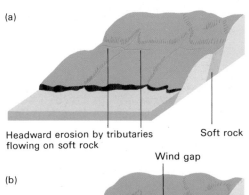

(a)

Headward erosion by tributaries flowing on soft rock

Soft rock

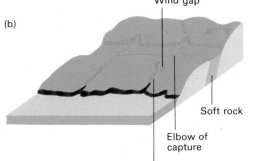

(b)

Wind gap

Soft rock

Elbow of capture

Misfit stream

When looking for an example of river capture, we should be able to suggest why one river has expanded at the expense of another and also look out for the features marked in Fig. 43. Where the capture took place one might expect to find a sharp bend or *elbow of capture*. Below this following the old stream course one would find a dry valley or *wind gap*, containing alluvium from the former stream. Eventually one would find a stream following the course of the old river but, because it no longer receives so many tributaries the stream is unusually small compared with the valley it is flowing in. It is a *misfit* stream.

This sequence of events has been identified in South Wales (Fig. 44) where the River Neath 'ate back' along a zone of weakness to capture both the Mellte and Hepste from the Cynon, leaving wind gaps to mark their former courses.

34 Study Figs 43 and 44. In the South Wales example,
  a Name a wind gap
  b Name a misfit stream
  c Which river is likely to have cut down vigorously because it has captured extra water from another river?

Fig 41 Incised meanders

Rejuvenation terrace

Incised meander

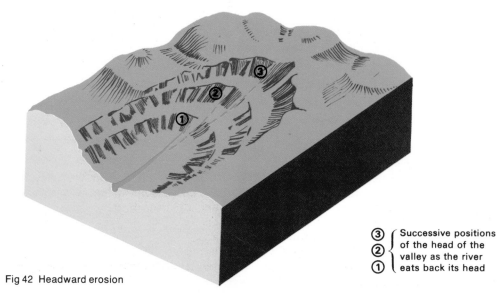

③) Successive positions
② } of the head of the
①) valley as the river
    eats back its head

Fig 42 Headward erosion

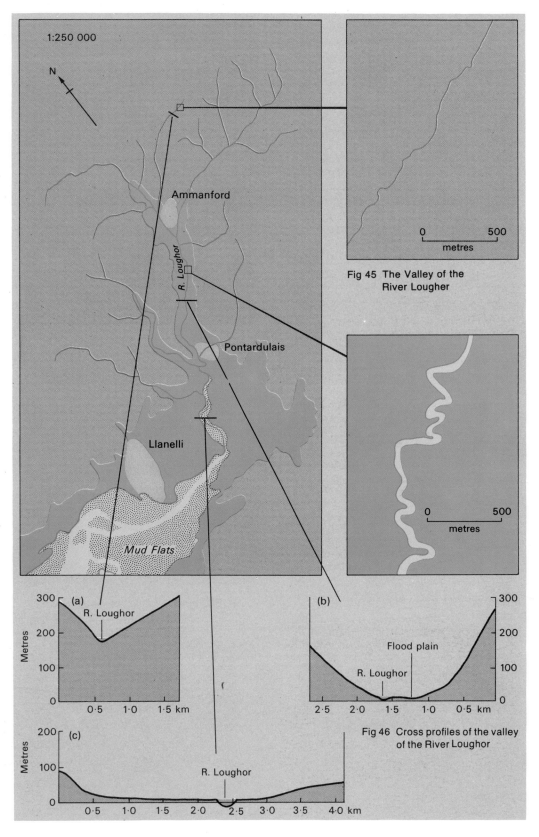

1:250 000

Fig 45 The Valley of the River Lougher

Fig 46 Cross profiles of the valley of the River Loughor

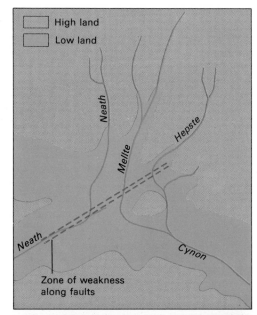

High land
Low land

Zone of weakness along faults

Fig 44 River capture by the River Neath, South Wales

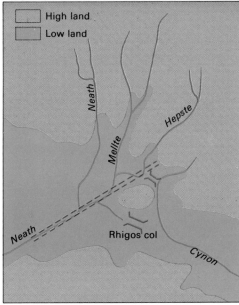

High land
Low land

Rhigos col

Case study

The River Loughor

The River Loughor (Fig. 45) flows into the sea near Llanelli in South Wales. Figs 47a, 47b and 47c are arranged in order from the river's source to its mouth.

61

35 *The river's long profile.* Draw a long profile of the river – that is, a diagram showing the slope from its source to its mouth. Use a vertical scale of 1 cm : 50 m and a horizontal scale of 1 cm : 1 km

Distance from source (km)	Height above sea level (m)
0	230
0.5	190
1	183
2	160
3	137
4	120
5	96
6	60
7	40
8	33
12	14
20	1

A more detailed diagram would show that the upper part of the river's course is not only steeper but also more irregular than the lower part. Use Figs 46 and 47 to describe the bed of the river in its upper course.

36 *The valley cross profile.* Study the three cross profiles of the Loughor's valley (Fig. 46). Describe how the valley's cross profile changes from source to mouth. Explain the changes you observe.

37 *The course of the river in plan* –that is, as seen from above.

a Use the enlargements of part of the lower course and part of the upper course of Fig. 45 to work out a Sinuosity Ratio for each length of the river.

b Draw a sketch map or series of sketch maps showing how you think meander A in Fig. 47 may change its shape during the next 5000 years.

(a)
(c)

(b)

Fig 47 Photographs of the River Loughor valley

Review Questions

1 Name examples of:
   a bird's foot delta
   a meandering river
   an area with a radial drainage pattern
   an example of river capture

2 Draw diagrams to show that you know the meaning of:
   a watershed   the water cycle   bluffs

3 How would you measure:
   the sinuosity of a river?
   the discharge of a river?

4 Explain: why a gorge often occurs just downstream of waterfalls.
   a Why the discharge of a river increases rapidly after a rainstorm but then takes longer to decrease to the normal flow.
   b What measures can be taken to prevent or control flooding.

6 Ice

Fig 1 Part of the Vatnajökull ice cap, Iceland

Look at Fig. 1. This photograph shows part of the Vatnajökull, an ice-covered mountain in Iceland. At the present time, large areas of the earth's surface are covered with ice.

1 Refer to Figs 1 and 2.
a Briefly describe the surface of the Vatnajökull ice cap.
b Work out the percentage of the earth's land area that is covered by ice today (Percentage of land covered by ice =

$$\frac{\text{area covered by ice}}{\text{total area of land}} \times \frac{100}{1})$$

c The surface areas of the earth that are more or less permanently covered by ice may be grouped into two types of location. What are they?

During the Ice Age 30 per cent of the surface of the earth was covered by ice – far more than today. Ice covered much of Britain and the landforms that existed before the ice came were changed by both the ice itself and its meltwater. Ice spread over Britain during the Pleistocene period of geological time (see Chapter 2, Fig. 10) and disappeared from Britain only about 10 000 years ago. Many of the landforms the ice has left may be clearly seen today and they form a significant part of the landscape of the British Isles. Geomorphologists are therefore very interested in the way the ice behaves and how it alters the surface of the earth.

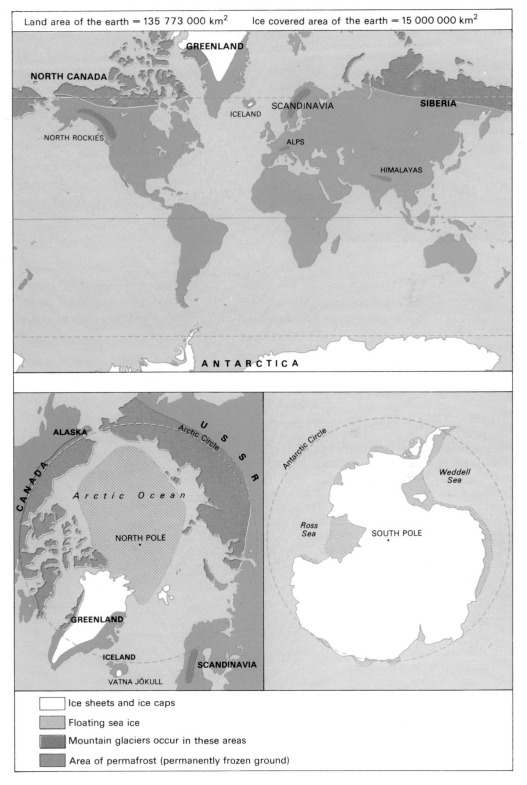

Fig 2 Ice on the surface of the earth

The Ice Age

Fig 3 Frost Fair on the River Thames 1683–84

Clearly, the climate was colder during the Ice Age than it is now, but the Ice Age itself was not one single, simple event. Climates are, in fact, always changing. Between about 1550 and 1850, for example, there was a 'Little Ice Age' when the climate grew steadily colder and 'Frost Fairs' were held on the frozen River Thames. The ice must have reached a considerable thickness to support the activities shown in the picture of the Frost Fair held on the River Thames in the winter of 1683.

Fig. 4 shows how the climate of Britain changed during and since the Ice Age. During this period ice advanced and retreated several times. The periods of ice advance are called *glacials* and the periods of ice retreat are known as *interglacials*.

2a What length of time, before the present, is shown on the graph?

b How many times did the ice advance over Britain during the Ice Age?

c When did the last glacial period start and how long did it last?

d Roughly how much difference in the July average temperatures is there between the glacials and interglacials? This figure is quite small.

e The diagram shows that the Ice Age was not just one period of very low temperatures, but a series of *comparatively* minor temperature fluctuations. Do you think we are still in the 'Ice Age'? Why?

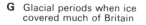

G Glacial periods when ice covered much of Britain

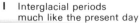

I Interglacial periods much like the present day

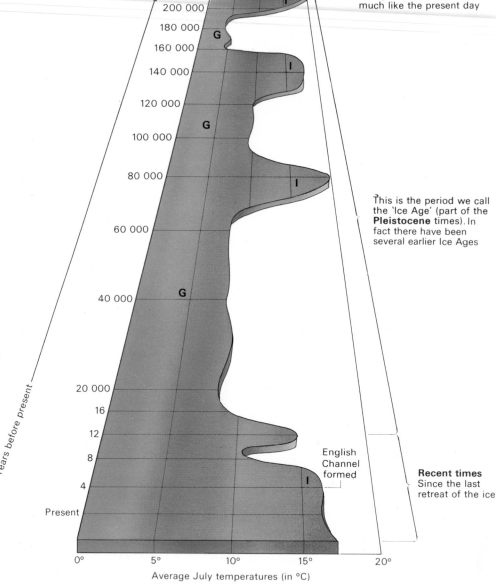

This is the period we call the 'Ice Age' (part of the **Pleistocene** times). In fact there have been several earlier Ice Ages

English Channel formed

Recent times Since the last retreat of the ice

Fig 4 Temperature changes in Britain over the last 200,000 years

During each of the glacial stages shown in Fig. 4 snow and ice accumulated in the upland areas. The ice then gradually spread further south. Most of the ice, in fact, originated in the higher land of Scandinavia, although there were local, high areas from which the ice advanced in Britain. Recent research also suggests that it may have moved eastwards up the Channel. The final shape of the British Isles was not yet visible at this time. Much of the North Sea was still land (for example the shallow Dogger Bank fishing grounds), and Britain was still joined to France. The extent of the ice over Britain is shown in Fig. 5. As the last glacial advance was not the furthest, the landscape produced by ice activity is complicated. Landforms are created by one advance of the ice and may be bulldozed by the next.

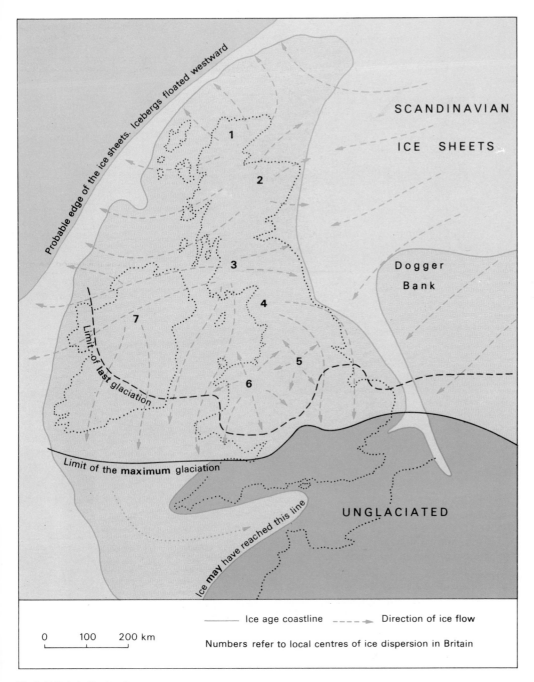

Fig 5 Britain in the Ice Age

Map labels:
SCANDINAVIAN ICE SHEETS
Dogger Bank
Probable edge of the ice sheets. Icebergs floated westward
Limit of last glaciation
Limit of the maximum glaciation
Ice may have reached this line
UNGLACIATED

Legend:
Ice age coastline ———
Direction of ice flow ----→
Numbers refer to local centres of ice dispersion in Britain

0   100   200 km

The accumulation of ice

In our present climate, any snow that falls in the winter melts during the spring. In highland areas, where the temperatures are lower, there is more snowfall and it takes longer to melt. If the climate were colder than it is now, the winter snowfall would not melt during the spring and summer months but continue over to the next winter when more snow would fall on top. In this way, snow accumulates on the surface of the land. Fig. 6 shows the changes that take place as each year's snow builds up. Air is removed as the ice is compressed. As the glacier ice becomes thicker it will move downslope by its own weight.

Fig 6 The build-up of ice

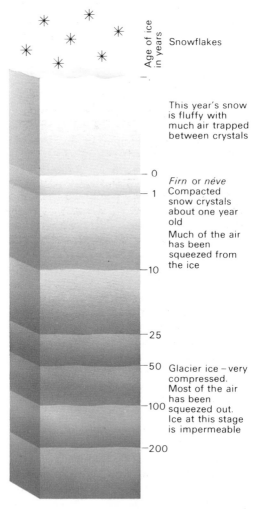

Age of ice in years

Snowflakes

This year's snow is fluffy with much air trapped between crystals

0
1 Firn or néve
Compacted snow crystals about one year old

Much of the air has been squeezed from the ice

10

25

50 Glacier ice – very compressed. Most of the air has been squeezed out. Ice at this stage is impermeable

100

200

3 Refer to Figs 4 and 5.
a Using an atlas, name the centres of ice dispersion in Britain (numbered 1–7 on the map, Fig. 5).
b When was the English Channel formed? What do you think caused the rise in sea level that brought about this separation of Britain from France?
c By referring to an atlas, describe (using placenames), the position of the line marking the limit of the last ice advance over Britain.

Ice movement

A mass of moving ice is called a *glacier*. It may move in a valley, like a river, or flow out from a mountainous area across a plain as a 'tongue' of ice (see Fig. 7). A very broad mass of ice flowing out from its centre in all directions is called an *ice sheet* (or, if smaller, an *ice cap* – like that shown in Fig. 1).

Fig 7 The Gorner glacier, Switzerland

A glacier cannot keep growing indefinitely. It will eventually move into warmer areas where the ice will melt. The length of the glacier may be divided into two parts:
i) the zone where the snow and ice are building up – *the accumulation zone*;
ii) the zone where ice is melting – *the ablation zone*.

Between these two zones is a line where the rate of accumulation is balanced by the rate of ablation. Fig. 8 shows how this line advances (as well as the snout of the glacier) in winter.

The balance between accumulation and ablation is known as the glacier's *budget*. If the rate of accumulation is greater than the rate of ablation then the snout of the glacier will move forward. If the rate of ablation is greater than the rate of accumulation then the snout of the glacier will retreat. Notice though that even if the snout of the glacier is retreating, the ice itself is still moving forward, because melting is taking place faster than the ice is advancing.

Although ice is solid it is able to flow. It does this in a combination of ways, depending on the temperature of the glacier:

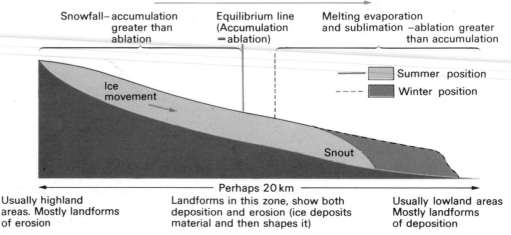

Fig 8 The glacier's budget

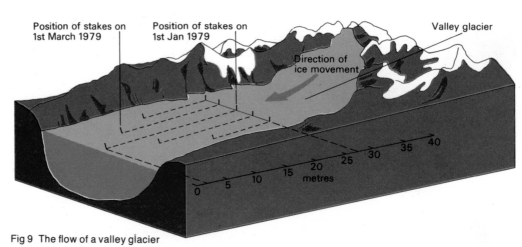

Fig 9 The flow of a valley glacier

i) by moulding its shape downhill in a plastic way, rather like thick treacle
ii) by partly melting on the upstream side of rocks in its path and then refreezing once it is past the obstacle
iii) by slipping over a thin layer of meltwater at its base
iv) by the movement of fairly solid slabs of ice past each other along lines of faults in the ice.

4 Refer to Fig. 9.
An expedition was made to determine the way in which this glacier flows. Stakes were driven into the ice in a straight line and their position recorded on a map. The position of the stakes was then examined two months later.
a Which stake has moved the furthest?
b How far has it moved?

c What is the average speed of this glacier in metres per day and metres per year?
The speed of ice varies within the glacier. Like water in a river (see p. 53) the fastest part is away from the sides and bottom.

The variation in the speed of the ice causes stresses within the glacier and cracks or *crevasses* develop. The speed of glacier flow varies within glaciers. It also varies *between* glaciers (from about 100 m to 7 km per year).

The geomorphologist's main concern with glaciers is the way in which they modify the surface of the earth. Like most other landform processes, ice modifies the earth by *erosion* and by *deposition*.

Erosion by ice

If the ice moves as slowly as in Fig. 9, where is the power for erosion? It is difficult to get underneath glaciers to see the process at work, so we are not sure precisely how glaciers erode. However, erosion seems to occur as a result of the combination of the following:

i) simple scooping up of material (including soil and loose rocks), that had been weathered before the Ice Age,

ii) the 'sandpaper effect' of rock particles trapped beneath the ice grinding away the solid bedrock. This process is known as *abrasion*. Deep grooves or *striations* are often left in the surface of the rock, like the scratches made by coarse sandpaper on smooth wood. Striations may be seen in areas from which the ice has only recently retreated,

iii) the removal of rock particles already loosened by jointing. Ice partly melts on the upstream side of rock obstacles in the path of the ice and refreezes on the downstream side. In this way, rock fragments are *plucked* away by the ice.

Fig 11 Glacial erosion and the transport of moraine

Fig 10 Striations on rocks after glaciation

These three processes may be seen in the section through a valley glacier in Fig. 11.

On the slopes above the surface of the glacier it is likely, in the cold climate, that the temperatures will be varying above and below freezing. There will be, therefore, much weathering by frost shattering and this loose material may fall onto the surface of the glacier and be transported away.

All material transported by the glacier is called *moraine*.

Fig 12 A roche moutonnée

5 Refer to Fig. 11.
a How were the following types of moraine formed?
lateral moraine
medial moraine
ground moraine
b Examine the landform in Fig. 12. Ice has moved over this mound of rock, passing smoothly over the upstream side and 'plucking' rock fragments from the downstream side.
Draw a sketch of this feature and show the direction of the ice. Label the diagram with a brief description of both sides.

This landform is called a *roche moutonnée*.

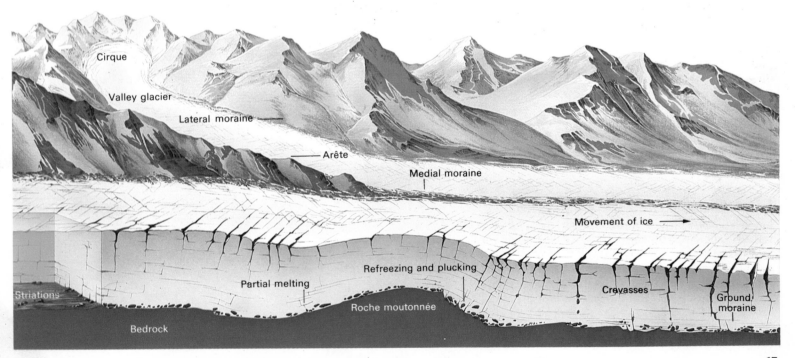

Landforms produced by glacial erosion

Most of these consist of features which existed before the ice age, but which have been enlarged or otherwise modified by the ice.

Cirques (or corries)

These are large, bowl-shaped hollows in highland areas and are generally the source of ice for glaciers. They vary in size but may be up to 2 km across. They are formed by the gradual accumulation of ice in a hollow in high land. Erosion takes place by freeze/thaw around the patch of ice and meltwater removes the eroded particles. This is called *nivation*. The ice gradually enlarges the hollow to the bowl, or arm-chair shape of a cirque. Two distinct features of the cirque may be seen after the ice has gone. These are the steep, rocky back wall (up to about 200 m high), and the 'lip'. The back wall was progressively cut steeper either by freeze/thaw activity between the ice and the rock face or by plucking away rock fragments broken by jointing. The lip of the cirque may be the result of the rotational slip of the ice. It is formed as the ice erodes less powerfully at the edge of the landform than the base.

As two neighbouring cirques become enlarged their back walls may meet 'back to back', resulting in a steep knife-edge ridge or *arête*. Where three of four cirques cut back on each other a *horn* or *pyramidal peak* may be formed at the centre.

Glacial troughs

Ice from the cirques moved down the valleys and joined to form glaciers. The valleys or troughs that were cut by the ice are broad and flat-bottomed, rather like the letter U in cross section. They are usually the result of glaciers enlarging pre-existing river valleys. As ice cannot flow so easily round corners as water the valleys are straightened out and the spurs of the

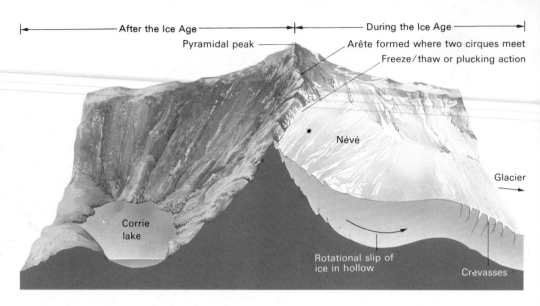

Fig 13 The formation of a cirque or corrie

river valleys are planed off or *truncated*. Rivers that flow in glacial troughs today appear too small for the size of the valleys cut by the ice. They are called *misfit streams*. Glacial troughs often contain long narrow lakes called *ribbon lakes*. These can be seen in the left hand photograph on p.1.

The larger troughs had more ice in them so cut deeper than their tributaries. These tributary valleys were left higher after the ice melted and are known as *hanging valleys*.

Look at the figure below. A major glacier once filled the large valley in the foreground. It was joined by a tributary glacier at A in the centre of the photograph. During glaciation the area would have looked rather like the scene in Fig. 1.

Fig 15 A hanging valley: the Upper Inn near St. Moritz

Fig 14 U-shaped glacial trough

6 a Explain the difference between the jagged outline of the mountains above the former level of the ice and the smoother shape of the valleys that were below the ice.

b A stream now flows down this *hanging valley*. What has happened to the stream at B?

c How was the landform at C formed? This feature is known as an *alluvial fan*. What might eventually happen if this process were continued? What use has man made of the alluvial fan? Why?

7 Look carefully at the photographs in Figs 14, 16, Fig. 1 on p. 2 and the left hand photograph on p. 1.

From the statements below, draw up two lists headed 'Characteristics of Glacial Troughs' and 'Characteristics of Cirques'.

> steep rocky back wall
> U shape in cross section
> lip which may hold back lake
> semi-circular shape
> may contain ribbon lake
> truncated spurs
> two cut back on each other to form an arete
> tributary valleys much higher.

Write the name of an example of each glacial feature at the end of each list.

Fig 17 High Street and Blea Water, Lake District

Fig 16 The Matterhorn – a pyramidal peak

8 Study Figs 17 and 18 carefully. The photograph shows High Street in the Lake District – an area which shows much evidence of glacial erosion. With the help of the map copy out and fill in the spaces in the following table:

Geomorphological name of landform	Place name of landform	Symbol on photograph
Cirque or corrie with corrie lake	Area around Blea Water	
	Upper Riggindale	●
U shaped valley		■
Arête		
Misfit stream		

69

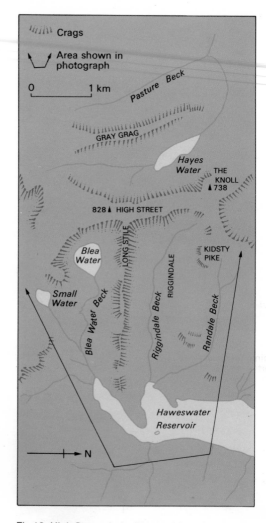

Fig 18 High Street, Lake District. Map of the area shown in Figure 17

Fiords

Fiords are drowned glacial troughs. The glaciers that eroded them reached the coast and the troughs were later drowned by the rise in sea level after the ice melted.

These deep inlets have a ridge or *threshold* at their mouths. This feature is much like the lip of a cirque. Most fiords are on the west coast of countries with high mountains not far inland. This means that when the ice was present the gradient of the glaciers was steep and the erosive power of the glaciers was great. Valleys that existed before the ice were therefore deepened. When the ice reached the sea, however, it melted and became thinner, icebergs broke away, and the erosive power of the glacier was much reduced at

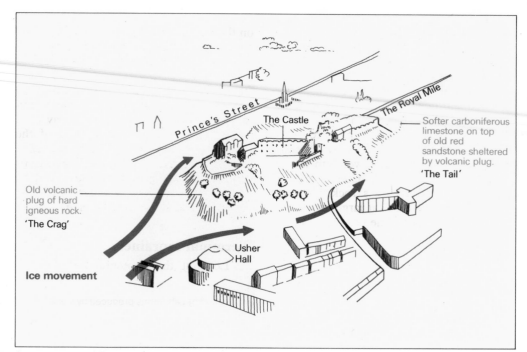

Fig 20 Edinburgh Castle and the Royal Mile: a crag and tail

Crag and tail features

9 Look at the diagram of Edinburgh Castle (Fig. 20).
Edinburgh Castle stands on a *volcanic plug*. How was this formed?
(Check with Chapter 3).
Why is it a good site for a castle?
Ice moved over this area from the direction shown by the arrow. Much Carboniferous Limestone was eroded Why do you think that the volcanic plug was left?

Fig 19 Nordfiord, Norway

the sea's edge. The present fiords, therefore, are deep inlets but nearer the sea they are comparatively shallow.

A strip of Carboniferous Limestone was protected by the volcanic plug and formed a smooth, moulded ridge. This now lies under the street known as the Royal Mile.

The complete feature, of volcanic plug and ridge of limestone, is known as a *crag and tail*.

Fig 21 The snout of a glacier

Deposition by ice

Landforms produced by glacial deposition are very complex. Melting ice deposits moraine mostly at the edges and ends of the glaciers. Later advances of ice bulldoze the deposits of previous advances, re-work them and pile them into new shapes. Deposition takes place in two main ways:

i) Melting ice dumps material on the surface of the land. This rock waste is *unsorted* which means that large and small rocks, as well as very fine particles will be jumbled together. This material may be further shaped by more ice passing over it. A covering of unsorted debris such as this on the landscape is known as *till* or *boulder clay*. Its precise composition varies according to the area over which the ice has travelled (and therefore the area from which the material has been eroded).

ii) Streams of meltwater flowing from the snout of the glacier carry away rock waste. This is deposited in front of the glacier by *river* processes. Deposits formed in this way tend to be *sorted*. This means that sands and gravels are arranged in layers as the streams have deposited first the larger and then the smaller particles.

Terminal moraines

10 Look at the photograph in Fig. 21 which shows a head-on view of the snout of a glacier. Describe the appearance of the ice and what is happening to it. The heap of debris in the foreground is terminal moraine. Refer to Figs 22 and 23 and explain how it has been formed.

Notice that the streams flowing from the snout of the glacier are carrying a large quantity of rock waste. This is dumped almost immediately as the streams are not powerful enough to carry it far. The river courses are therefore split up into many channels. This is known as *braiding*. Braided streams are very common on the margins of ice sheets.

Fig 22 and 23 a) Deposition by ice and meltwater (background) and b) Landforms produced by ice and meltwater deposition (foreground)

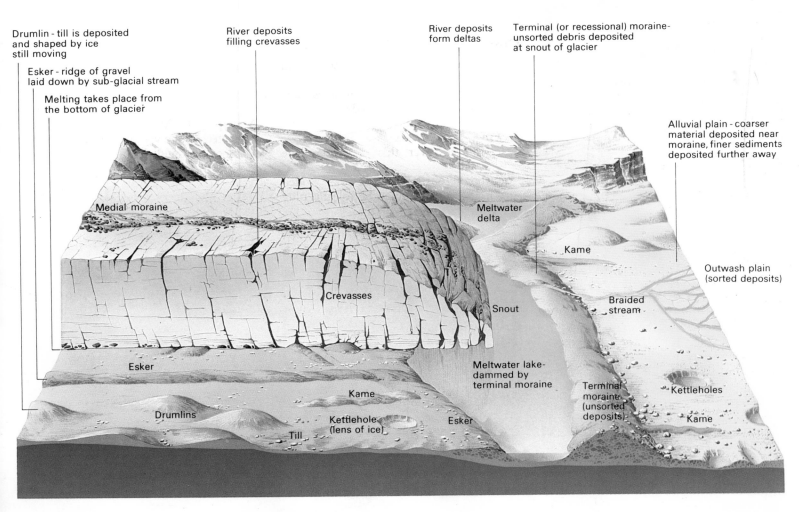

Drumlin - till is deposited and shaped by ice still moving

Esker - ridge of gravel laid down by sub-glacial stream

Melting takes place from the bottom of glacier

River deposits filling crevasses

River deposits form deltas

Terminal (or recessional) moraine - unsorted debris deposited at snout of glacier

Alluvial plain - coarser material deposited near moraine, finer sediments deposited further away

Medial moraine

Meltwater delta

Kame

Crevasses

Snout

Outwash plain (sorted deposits)

Braided stream

Esker

Meltwater lake - dammed by terminal moraine

Terminal moraine (unsorted deposits)

Kettleholes

Drumlins

Kame

Till

Kettlehole (lens of ice)

Esker

Kame

Drumlins

11 Describe the surface of the land in Fig. 24. Each of the hummocks is a *drumlin*. They are formed from boulder clay which has been deposited by the ice but shaped while the ice was still moving. The end facing the ice is more blunt than the other, tapering end. Can you see from the photograph which end is which? The ice moved from left to right.

Drumlins rarely occur singly but usually in *swarms*.

12 Study the aerial photograph of the drumlin swarm at Strangford Lough, Co. Down.

a What has happened since the drumlins were deposited?

b Locate Strangford Lough with the help of an atlas. Refer to Fig. 5.
Which direction did the ice that deposited these drumlins come from?

c Do you think that drumlins are landforms of glacial erosion, or deposition, or both? Why?

Erratics

13 Examine the photograph, Fig. 26, and the geological time scale and geological map (pp. 10–11). The geologist is examining a boulder of Silurian rock resting on Carboniferous Limestone.

a This boulder has been transported to its present position. Why could it not be the weathered remains of a layer of rock previously above the limestone?

b This rock has not been transported by man. What is the only possible natural method of transport?

Fig 25 Strangford Lough, Co. Down. A drumlin swarm

Fig 26 Dark Silurian boulder resting on light Carboniferous limestone

Fig 24 Drumlins: Glen Varragill, Isle of Skye

c Rocks transported in this way are called *erratics*. How could they be of use to geomorphologists investigating the flow of ice across Britain during the ice age?

d Refer to the section in Chapter 4 on limestone. Suggest why the limestone has not been weathered so much beneath the erratic.

Early man thought that there was something mysterious about these rocks. There are many legends associated with erratics.

Fluvioglacial landforms

Outwash plains

Melting ice obviously produces a great deal of water. Fig. 21 shows this flowing from the glacier snout as streams. These streams carry away the moraine deposited by the ice and deposit it a second time in front of the glacier snout. As these second deposits are laid down by water, however, they are *sorted*. Finer material is carried further away from the margin of the ice. Depending on the number of streams and the relief of the area in front of the ice edge, these deposits may build up a large *outwash plain*.

Eskers

Fig. 27 shows a long ridge of deposited material. This *esker* runs parallel to the direction of ice flow. Eskers are formed by deposition from streams flowing under the ice (see Figs 22 and 23)

Fig 27 An esker

Kames

These are made up of sorted debris which has been washed in to crevasses in the stagnant ice. As the ice melts, this debris is dumped on the ground. They may also be formed by the build-up of deltas in lakes at the ice margin. Fig. 28 shows a photograph of a kame in the Glaven valley in Norfolk.

Kame terraces

When lakes at the margin of the ice become completely filled with sediments a terrace may be left when the ice retreats. They may also be formed by deposition along a stream following the edge of the glacier.

Fig 28 A wooded kame in the Glaven Valley, Norfolk

Glacial diversion of drainage

Geomorphologists have to do a great deal of detective work to investigate some landforms. Consider, for example, the mysterious case of the rivers and valleys of the North Yorkshire Moors ... Follow the course of the River Derwent on an atlas map of Britain. The river rises about 15 km to the northwest of Scarborough. It flows southeast, parallel to the coast for about 20 km. It then ignores a short lowland route eastward to the coast but follows instead a detour of about 90 km through high land, cutting deep gorges as it does so.

Next, look at the photograph below. This shows a large, deep steep-sided valley. The river in this valley is very small – far too small to have cut a valley of this size. This is Newtondale, cut into the south side of the North Yorkshire Moors.

Fig 29 Newtondale, North Yorkshire Moors

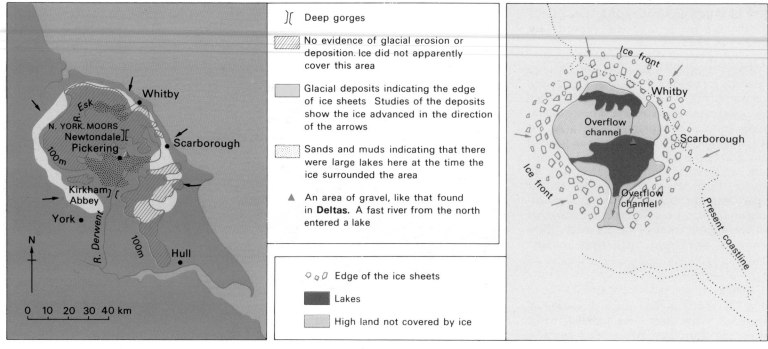

Fig 30 The North Yorkshire Moors

Legend (Fig 31):

](Deep gorges

No evidence of glacial erosion or deposition. Ice did not apparently cover this area

Glacial deposits indicating the edge of ice sheets Studies of the deposits show the ice advanced in the direction of the arrows

Sands and muds indicating that there were large lakes here at the time the ice surrounded the area

▲ An area of gravel, like that found in **Deltas**. A fast river from the north entered a lake

○ ₀ ◊ Edge of the ice sheets

Lakes

High land not covered by ice

Fig 31 The formation of glacial lakes and overflow channels

To explain these, and other puzzles of the North Yorkshire Moors, geomorphologists have carefully gathered evidence of past events. Much of the evidence is presented on the map in Fig. 30. Study the clues carefully. Can you discover what has happened to cause the diversion of the River Derwent and the formation of Newtondale?

Here is a possible solution. Ice surrounded the Moors, blocking the exit of the River Esk. The water was ponded back to form a large lake known as Lake Eskdale. This lake overflowed, its fast flowing water cutting the gorge at Newtondale and depositing the delta at Pickering. This overflow water, together with the water from the River Derwent created a second large lake – Lake Pickering – also ponded back by ice from the east. This lake overflowed to the west, cutting the gorge at Kirkham Abbey. After the ice retreated the River Esk followed its old course to the sea but the River Derwent continued to follow the course diverted by the ice.

Periglacial processes and landforms

Fig. 2 on p. 63 shows the major area of *permafrost* or permanently frozen ground in the world today.

14a What does 'peri' at the beginning of a word mean? (Look this up in a dictionary).
 b Describe where the area of permafrost is in relation to the ice sheets.
 c Refer to Fig. 5 on p. 65. Which areas of the British Isles would have been at the edge of the ice sheets for much of the Ice Age?

Periglacial processes are those which act *near* or *around* the ice sheets. One of the major features of these areas is that the ground is permanently frozen – sometimes to 300 m deep.

At depth the permafrost does not thaw but the top one or two metres of this frozen ground may melt in summer and re-freeze in winter. This layer is known as the *active layer*.

Periglacial processes

Many processes around the ice edge may be explained by the way water behaves at low temperatures. The expansion of water when it freezes causes frost shattering of rocks. Water trapped beneath the ground, on freezing, causes the soil and surface layers of rock to heave upwards. Nivation (the process responsible for the early development of cirques) is also common under periglacial conditions.

Imagine a gentle slope under periglacial conditions. The surface, active layer is frozen in winter and thaws in summer. When the active layer melts the surplus water cannot soak downwards as the layer beneath is frozen. The whole active layer is made heavier as it becomes waterlogged. Under these conditions the surface layer begins to move downslope. There are many areas in Britain where you can see rock fragments which accumulated at the foot of slopes under periglacial conditions during the Ice Age. These deposits are known as *head*.

Wind action is also common in periglacial areas. Strong winds often develop as there is so little vegetation (because of the cold) to act as wind breaks. Fine particles of dust and silt are picked up. This material has been deposited in areas of Northern Europe, dating from the last Ice Age and is known as *loess*.

Periglacial landforms

One of the most striking effects of frost in periglacial areas is *patterned ground*. There are different types of patterned ground. This example shows stones arranged in various polygons.

In winter, as ice grows in the active layer it expands and pushes stones towards the surface. Stony areas drain water away more effectively and therefore stay relatively dry. Less water collects to freeze and therefore there is less upward heaving of the ground. The muddy areas between are moist as water is trapped in the spaces between the mud and the small stones. On freezing these areas are pushed up into small domes and surface stones roll to the side. In this way a pattern of stone circles or polygons is formed.

Fig 32 The formation of patterned ground

Fig 33 Large pingo, Northern Alaska

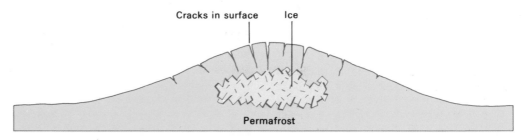

Fig 34 The formation of a pingo

Fig. 34 shows a mass of water which has become trapped beneath the surface in a periglacial environment. As temperatures fall the growing lens of ice exerts pressure on the surface and a dome-shaped hill is formed. These hills are known by their eskimo name of *pingo*. Sometimes they reach very large sizes – up to 50 m in height and 500 m in diameter. The ice may bulge up so much that the surface cracks. The ice inside is exposed and usually melts.

15 Examine the photograph of the pingo (Fig. 33). Describe its shape and explain how it was formed.

Review Questions

1 Describe briefly three ways in which geomorphologists can discover the direction of ice flow in glaciated area.
2 Look again at the photographs of the landforms of glacial erosion and glacial deposition in Britain. Why are landforms of glacial deposition not so well preserved as those of glacial erosion?
3 Look at each of the photographs of the landforms of glacial deposition in Figs 21, 24, 27 and 28, and for each say whether you think the material they are made of is sorted or unsorted.

7 Coasts

The action of the sea

The power of nature to shape the land is very clearly seen along our coastline. Inland most features alter so slowly that the change is hardly noticed but this is not so along the coast. On some stretches of coast landslips happen every year, while there are daily variations in the shape of many beaches.

Waves

Waves are the cause of most of these changes. They form as a result of the wind blowing over the sea and ruffling it up. The stronger the wind is, the bigger the waves that result. Large waves also occur when the wind has blown for a long distance over the sea. In the same way, if you look at a pond on a windy day you will notice that the waves gradually increase in size as they move across the surface. The further they are from the up-wind side of the pond, the larger they become. The length of water over which the wind has blown is called the *fetch*.

1 Plot a graph to show how the fetch influences the size of waves by completing Fig. 2 using this information:

FETCH (km)	WAVE HEIGHT (metres)
50	2.5
100	3.5
150	4.2
200	4.6
250	4.9
300	5.1

These figures assume that the wind is blowing at 10 km an hour. As the waves will only build up to this size over a period of time, the figures also assume that the wind has been blowing for at least twenty hours.

Fig 1 Start Bay, South Devon

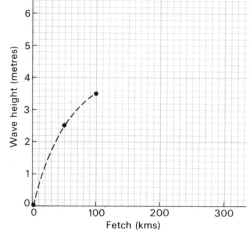

Fig 2 Axes for plotting fetch against wave height

2 Use an atlas map of Britain to find the location of Whitby and Dover. Then, using a map of Western Europe, measure the distance of fetch.
a at Whitby from the north and south-east.
b at Dover from the south-east and from the east.
Which place is more exposed to large waves?

At Whitby one would expect the largest waves to come from the north because of the long fetch in that direction. The largest waves at a place are called the *dominant* waves and we shall see that these are very important for moving sand and pebbles along the coastline.

The behaviour of waves out at sea may be compared with the waves which pass along a rope if you hold one end and flick it up and down. The shape of a wave moves along the rope but the rope itself does not move away from you. In the sea it is the shape of the wave that is moving forward rather than the sea itself. If you watch a piece of wood floating in deep water you will notice that it is not carried forward by the wave. It just bobs up and down with slight forwards and backwards movements as the wave passes under it.

When waves approach the shore however, they are affected by the sea floor and behave in a different way. In shallow areas they move more slowly than they do in deep water. This has an important result which is known as *wave refraction*. Study Fig. 3 which shows a wave approaching the shore. At A it is straight, but soon after this it runs into shallow water off the headlands. It slows down there but keeps moving forwards at its original speed into the bay. So when the wave reaches B it is no longer straight. Indeed by the time it reaches the land it is approaching almost directly onto the coast wherever you are. You may have noticed the results of wave refraction when you have been at the seaside. It is very unusual for waves to travel along a beach in the way shown by Fig. 4. Instead, the waves are bent round as they come into the shallow water (Fig. 5) so that they seem to be coming from almost straight out to sea.

3 Make a tracing of the coastline shown in Fig. 3. Draw in the position of a wave at three times assuming it was coming from the southwest rather than from the south.

Fig 3 Wave refraction

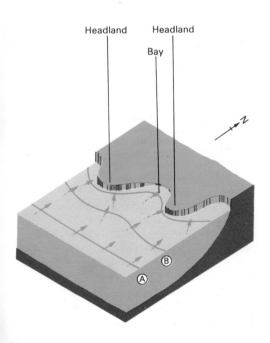

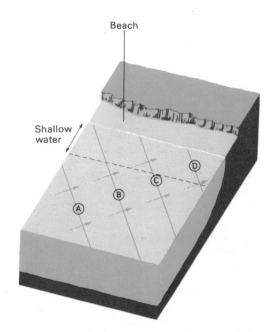

Fig 4 It is unlikely that waves would move along a beach like this

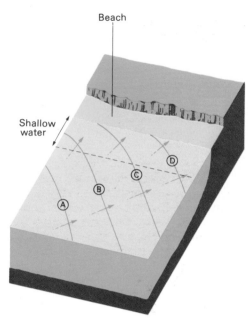

Fig 5 It is more likely they would be bent like this

When a wave moves into really shallow water the sea floor interferes with its movement. The wave then topples over or *breaks*. This happens when the depth of the sea is about the same as the height of the wave. From this moment on it is no longer only the shape of the wave that is moving forwards but the wave itself, as any surf boarding enthusiast will tell you. Most of the erosion and transport of material along the coast is carried out by the massive power of waves when they have broken.

Tides and currents

The influence of waves is greatly increased by the effect of the *tide*. The level of the sea moves up and down once every 12½ hours. This means that the waves can act on the coast for up to five metres above and below the average (mean) sea level. Tides are caused by the pull of the moon's gravity. The sun's gravity also has a small effect so when both sun and moon pull in the same direction the high tides are higher and the low tides lower than usual. These are called *spring tides*. When the sun and moon are pulling in different directions the range between high and low tides is less (*neap tides*).

Tides are important for allowing the sea to attack a greater area of the coastline because of the vertical movement of the sea every day. They are also the major cause of *currents* around our coasts. These are areas of moving water which may be very wide and slow like the Gulf Stream and North Atlantic Drift – the results of winds rather than tides – or much narrower and more local. They are of great importance to sailors who can be carried many kilometres off course if they do not know the way the currents move. They are less important than waves to the geomorphologist for they mainly carry fine mud rather than the sand and pebbles that waves can move. But their influence should not be forgotten in estuaries and other narrow passages along our coast where vast currents of water move in and out with the tide. For example, a tidal current flows through the Menai Straits (between Anglesey and the mainland of North Wales) at a speed of 15 km/hour, and fast currents are also found at the mouths of the Teign and Dart estuaries in South Devon (Fig. 6). These currents can result in daily changes to the shape of mud flats – a matter of interest to both geomorphologists and sailors.

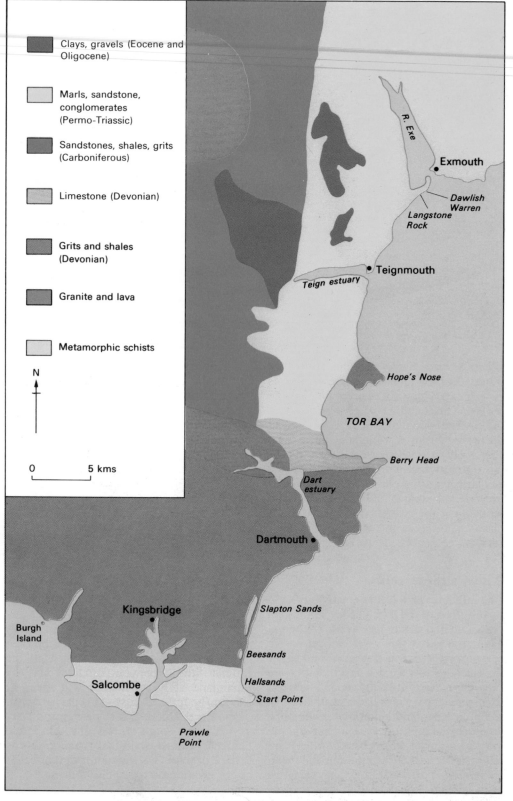

Fig 6 Map of the South Devon coast

The legend for the map reads:

Clays, gravels (Eocene and Oligocene)

Marls, sandstone, conglomerates (Permo-Triassic)

Sandstones, shales, grits (Carboniferous)

Limestone (Devonian)

Grits and shales (Devonian)

Granite and lava

Metamorphic schists

N

0   5 kms

The rest of this chapter looks at the way in which waves and currents shape the coastline. We shall particularly study the features of the South Devon coast.

4 Study Fig. 6 and an atlas map showing the English Channel.
a From which direction will dominant waves approach the coast between Start Point and Exmouth?
b Revise (Chapter 2) the nature of the rocks marked on Fig. 6.

Erosion

How the sea erodes

The sea erodes the land by *abrasion*, *hydraulic action* and *solution*. On a stormy day the crash of stones as they are hurled against the rocks shows the power of large waves. At such times the stones themselves are rounded by the process of *attrition*.

5 Revise the meaning of the terms *abrasion*, *hydraulic action*, *solution* and *attrition* (p. 54).

Another powerful form of erosion occurs when waves break against a cliff where air is trapped in joints or hollows. The air is greatly compressed and this enlarges the crack. When the wave retreats the air expands with explosive force weakening the rock still further.

The results of erosion

Erosion is most important where the land is exposed to powerful waves. These cut a *notch* in the rocks near to high tide level (Fig. 7) and as this is enlarged a steep rock face or *cliff* will be formed if the sea is eroding a highland area. Steep cliffs often occur where erosion is fast because in such places the sea undermines the cliff rapidly, forming a large notch which leads to the collapse of the cliff above. Where the sea erodes less actively, the cliff face is worn back by weathering and becomes more gentle.

Fig 7 Waves cut a notch in the rocks near high tide level

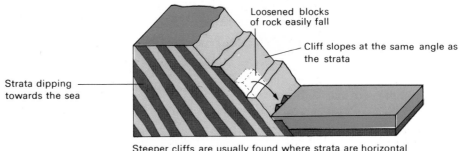

Loosened blocks of rock easily fall

Cliff slopes at the same angle as the strata

Strata dipping towards the sea

Steeper cliffs are usually found where strata are horizontal or dipping inland

The type of cliff depends on the geology of the coast as well as on the speed of erosion. Impressive cliffs are generally found in areas of hard rock or in places where the rock contains vertical joints. The famous 'White Cliffs of Dover' are chalk cliffs following vertical joints in an area of active erosion. The angle of dip of the rocks may also influence the steepness of the cliff.

In areas where soft rocks are being eroded, the collapse of cliffs into the sea may result in the loss of many hectares of land each year. Farmland, footpaths and even buildings are destroyed in this way. On the Holderness coast of Yorkshire where there are boulder clay cliffs, about two metres of land a year are being eroded away. This amounts to the disappearance of several kilometres of land since Roman times, including not only farmland but entire villages.

Where cliffs are 'retreating' in this way, the rocks which were previously underneath the cliffs are exposed to the sea and eroded by it into a *wave cut platform*. Fig. 10, taken at low tide, shows a large wave cut platform below chalk cliffs at Flamborough, Yorkshire.

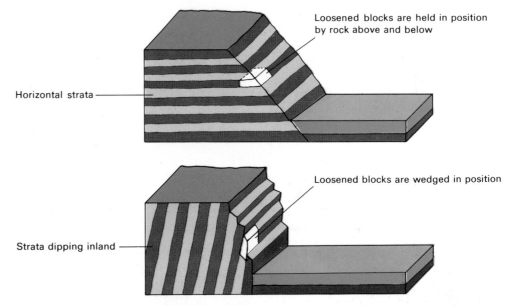

Loosened blocks are held in position by rock above and below

Horizontal strata

Loosened blocks are wedged in position

Strata dipping inland

Fig 8 The steepness of cliffs may be influenced by the angle of dip of the rocks

Fig 9 The formation of a wave cut platform

Fig 10 Wave cut platform at Flamborough, North Yorkshire

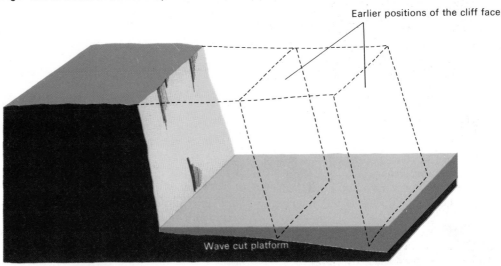

Earlier positions of the cliff face

Wave cut platform

Inspection of a cliff face often reveals features which result from the erosion of lines or zones of weakness in the rocks. The sea is particularly good at picking these out. Where the waves enlarge a joint or fault a *cave* may form. Air is then trapped and compressed inside the cave and may force its way up cracks to the cliff top. If you were standing at such a point on the cliff top you could put your hand by the crack and feel a blast of air and spray coming out each time a wave crashed into the cave below. Such a feature is called a *blow hole*. Fig. 11 shows a blow hole at Flamborough Head, much enlarged by the collapse of part of the cave roof.

If a cave grows in length across a narrow headland it may eventually extend all the way through or meet with another cave on the other side. When this happens the cave, now open to the air at both ends, is known as an *arch*. Fig. 12 shows an arch at Langstone Rock, Devon. Notice how the arch has formed where erosion has picked out the line of a fault.

At a later stage, if the roof of an arch collapses, part of the narrow headland becomes a small island, known as a *stack* (Fig. 13). Some stacks are spectacular, high rocks while others have been worn away for many years and can only just be seen above the sea.

Bays and headlands

When the sea erodes a stretch of coastline it not only picks out the joints and small scale weaknesses in the rocks but also attacks weak areas on a larger scale. If we look at the coastline shown in Fig. 6 we notice that there are several *headlands* – areas extending out into the sea – and between these are *bays* where more of the land has been worn away. Headlands tend to be formed where there are harder rocks which resist erosion by the sea.

6a Name the headlands in S. Devon where the coast is formed of resistant granite and lavas.
 b Explain why a bay has been formed between Hope's Nose and Berry Head.

This type of coastline with distinct bays and headlands often occurs where bands of hard and soft rock meet the coast more or

Fig 11 Blowhole at Flamborough, North Yorkshire

Fig 12 An arch. Langstone Rock, Devon. The arrow indicates the fault line

Fig 13 A stack: Freshwater Bay, Isle of Wight

less at right angles. It is known as a *discordant* coastline. Discord means disagreement and the coast is given this name because it cuts across the line of the rocks rather than following them. Where rocks and coastline lie parallel to each other the coast tends to be much straighter and is known as a *concordant* coastline, concord meaning agreement.

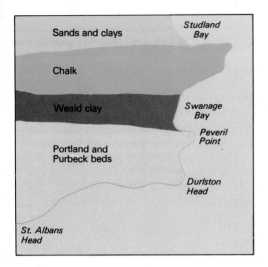

Fig 14 A discordant coast, Dorset

7 Study Fig. 14.
a Which stretch of coast is concordant and which is discordant?
b Look at the pattern of bays and headlands. Which rocks do you think are hard and which are soft?

Erosion and people

Where beaches or cliffs are being rapidly eroded, man is likely to be very concerned especially where farm land, holiday beaches or houses are involved. So in several parts of the country large sums of money are spent on *sea defences*. Sometimes a *sea wall* – a concrete cliff – is built to protect the land from erosion. Another way of protecting the coast is to build *groynes*, wooden fences running down a beach into the sea. These keep the sand or shingle in place and by maintaining a larger beach, prevent waves from attacking the cliffs behind.

Fig 15 Groynes protect the coast

Fig 16 Derelict cottages at Hallsands, Devon

It is very easy to upset the balance between erosion and deposition along a stretch of coastline. Fig. 16 shows derelict fishermens' cottages at Hallsands, Devon. In the nineteenth century the men used to keep their boats on a beach which lay between their houses and the sea. But in 1897 650 000 tonnes of shingle were dredged from the sea bed a few kilometres out to sea. This shingle was used to extend the naval dockyard at Devonport, but its removal resulted in waves attacking the beach more vigorously. In a few years it had disappeared and the waves then proceeded to erode the rocks on which the houses were built. Within twenty years the village was abandoned.

Transport and deposition

Longshore Drift

Water moving up the beach after a wave has broken is known as *swash*. When waves are breaking at an angle to the beach the swash moves not only up the beach but also slightly along it. This water then runs straight down the slope of the beach into the sea. It is then known as the *backwash*. So a pebble picked up at the water's edge by the swash and returned by the backwash would end up a short distance along the beach. This movement of material, which occurs on any but the calmest days, is called *longshore drift*.

8 An exercise to demonstrate longshore drift. If you are able to visit the coast, place a number of small, brightly painted pebbles at the water's edge. Note whether the waves are approaching the beach from straight out to sea or at an angle. See whether the pebbles are moved along the beach in one direction or whether they are scattered at random in both directions. Explain the movement of the pebbles in terms of the wave direction you noted.

Fig 17 Longshore drift

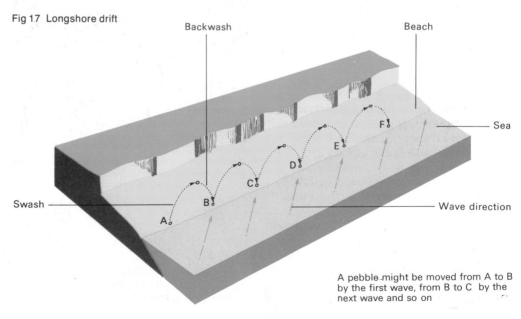

A pebble might be moved from A to B by the first wave, from B to C by the next wave and so on

Suggestions:

a a large number of pebbles is needed for this practical as some will be dragged out to sea or be buried in the beach.

b Remember that large pebbles will not be moved if the waves are small.

c It is a good idea to pick up the pebbles at the end of the practical and take them away with you, leaving the beach as tidy as possible.

The direction of longshore drift on any one day is decided by the direction of the waves approaching the shore that day. But if a beach is studied for a period of several years, it is clear that there is an overall movement of material in one particular direction. In Fig. 15 you will notice that there is more material on the right hand side of the groyne than there is to the left. This is because longshore drift is generally moving the pebbles from right to left and they have built up on the right hand side of the groyne. The overall direction of longshore drift is determined by the direction of the dominant waves approaching the shore.

9 Revise the idea of *fetch* and how this determines the direction of the *dominant waves*. Then study an atlas map of the British Isles and work out the direction of longshore drift you would expect to find

a in E. Anglia between Yarmouth and the Thames Estuary

b along most of the south coast of England.

Spits, bars and tombolos

The movement of material along the coast by longshore drift is interrupted where an estuary cuts across the coastline. Material gathers at the side of the estuary and over the years this area of deposition grows in length as fresh material is added to it by longshore drift. Gradually it is extended outwards into the estuary, forming a *spit* – a finger-like area of deposited sand or shingle. Some material is removed from the end of the spit by waves and tidal currents. The spit grows when the input of material by longshore drift is greater than the amount removed from it. When the

input (p. 4) and output are the same, the spit stops growing. Many spits continue the direction of the coastline across a bay or estuary in this way, though sometimes they leave the coast at a slight angle. This is because they tend to form at 90° to the direction from which the dominant waves are coming.

Fig. 18 shows Dawlish Warren, a spit at the mouth of the Exe estuary. It is two kilometres in length and has several features typical of spits round the British coast. Notice how it curves round at the bottom of the photograph. Here, at a distance from the land, the sand is more easily forced inland by the waves.

Behind Dawlish Warren lies an area of *salt marsh*. Mud is deposited in this area, which is protected from the large waves of the open sea by the spit. As the water becomes shallower mud is exposed at low tide and special types of marsh vegetation start to grow. Eventually the area is only flooded at spring tides or during storms and may be used as pasture land. Salt marsh is often broken up by small muddy creeks which are occupied by water at high tide.

On the spit itself, sand blown up from the beach has gathered as *sand dunes* – hills of sand held in position by special long-rooted grasses such as *marram grass*. Dunes which have no grass growing on them change shape according to the strength and direction of the wind, so in many such areas marram grass has been deliberately planted to prevent erosion of the dunes.

Spits often change shape over short periods of time. One storm can completely alter the end of a spit or even break through it where it is narrow. There are also longer term changes, sometimes the results of man's activities. At Dawlish Warren the spit has been badly eroded over the last hundred years. It is thought that the railway built along the coast south of Dawlish in 1846 has protected the cliffs from erosion. Less material has therefore been supplied to the spit by longshore drift – in other words, the input has decreased. As a result more material has been eroded than has been added to the spit.

Fig 18 Dawlish Warren, a spit at the mouth of the River Exe, Devon

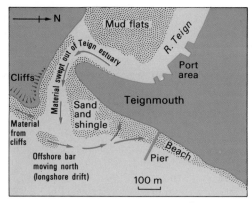

Fig 19 An offshore bar at Teignmouth, Devon

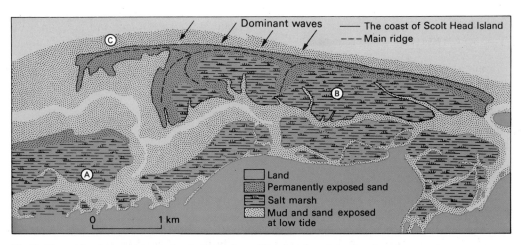

Fig 20 Map of Scolt Head Island showing laterals and the direction of dominant waves

Another depositional feature, eight km further south along the Devon coast is shown above. It is an *offshore bar*, consisting of a sand bank built up from material carried into the sea by the River Teign and moved along the coast by longshore drift. As it is exposed above the sea only at low tide and is constantly changing its position it is a hazard to ships entering the Teign estuary. In some parts of the country there are larger offshore bars, exposed at all stages of the tide. One of the largest is Scolt Head Island off the north Norfolk coast.

10 Study Fig. 20

a Explain why salt marsh has grown up at A and B but not at C.

b In which direction does longshore drift move material along the bar?

c 'Scolt Head Island has grown in a series of spurts. When its length has remained the same for some time its seaward end has been forced to curve round inland.' What evidence on the map might be used to support this statement?

A *baymouth bar* is a finger of deposited material which, unlike a spit, completely crosses a bay from one side to the other. Slapton Sands in Devon is a well known example (Fig. 21). The bar is part of a shingle beach which extends seven km northwards from Hallsands, crossing the mouths of several bays. A large lake – Slapton Ley – lies behind the main bar. It has been partly filled in with mud deposited by streams entering the northern end of the Ley. Some doubt has been expressed about how Slapton Sands

Fig 21 Slapton Sands

formed. Is it a spit that has steadily grown longer until it touches the land at both ends of the bay? Or did it start as an offshore bar, forced inland by waves until it became the feature we see today? An investigation into the origins of the shingle in the bar suggests that the second idea is probably the correct one.

A similar feature to the baymouth bar is the *tombolo*, a depositional landform connecting an island to the mainland. The most famous example in Britain is Chesil Beach and in Fig. 22 you can see the similarity between this feature and Slapton Sands. Both are long, narrow shingle ridges with a lagoon of calm water between the ridge and the former coastline. Rather more common are small tombolos only a hundred metres or so in length such as that

at Bigbury-on-Sea, Devon. Burgh Island may be reached from the mainland at low tide when the tombolo is exposed above sea level (Fig. 23).

Fig 22 Chesil Beach

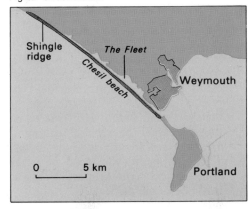

Fig 23 a) A photograph of Burgh Island

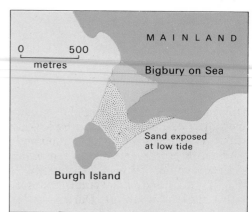

b) Map of Burgh Island

Beaches

The depositional feature which is most familiar to us is the *beach*. The largest beaches are usually found in sheltered bays rather than at headlands where erosion is more common than deposition. The size of a beach from back to front is determined by its gradient. The widest beaches often consist of fine mud and these slope very gently into the sea. Sandy beaches usually slope more gently than shingle beaches so as a general rule we can say that the larger the material making up a beach, the steeper it tends to be. Sometimes this is clearly seen on a single beach. There may be a steep ridge at the top of the beach consisting of stones thrown up by storms, especially at spring tides. Lower down the beach more gently sloping sand may be exposed as the tide goes out.

11 An exercise to investigate beach profiles.

If you are visiting a beach, as well as studying longshore drift (p. 81) you could also see whether the ideas expressed in the last paragraph apply to that particular beach. Is it true to say that 'the larger the material making up a beach, the steeper it tends to be'? The equipment you need is a clinometer and some calipers. Both can easily be home-made.

a Select a number of places on the beach which have different gradients and consist of different materials. (Or you may have to visit several beaches to get sufficient variety.)

b At each place measure (i) the steepest gradient (ii) the average size of the material. If the particles are too small to be measured, class them as mud, sand or shingle.

c Draw a graph similar to Fig. 23 (c) and see whether our statement about beaches is true.

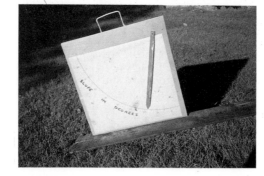

Fig 24 a) Clinometer

b) Calipers

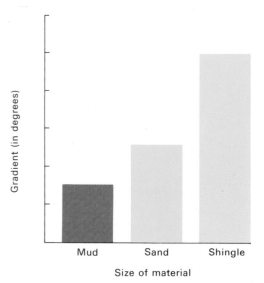

c) Graph of Beach gradients

Features such as ridges on a beach are caused by the work of waves. A change in the direction or size of the waves results in changes to the shape of the beach. The beach and the waves are part of a *system* (p. 4) and if one part of the system changes (or gets out of balance), the rest of the system will change in response to it. So on some days waves will build up the beach while on others they will drag material away down to the sea. Ridges on the beach may appear and disappear in only a few days. These changes in gradient from day to day make beaches a fascinating place for the student of landforms as well as for the holidaymaker.

Changes of sea level

There are some features around our coastline which we can only explain by looking back over thousands or tens of thousands of years. In particular we must remember that sea level has changed many times in the Pliocene and Pleistocene periods, sometimes by over 100 m. During the Ice Age for example, sea level was low in the glacial periods but high in the inter-glacials (see p. 64). What have been the results of these changes?

Emergence

A fall in sea level results in the shallow floor of the sea emerging from under the water and becoming dry land. A rise in the level of the land has the same effect. Much of Scotland is now rising by several millimetres a year so the land has risen by several metres since the last glacial period. Beaches and cliffs formed in the past have therefore been lifted well above sea level. On Fig. 25 you can see a gently sloping, grass covered *raised beach* backed by an old *cliff line* above the sandy beach and cliffs at the present sea level.

Submergence

If the sea level rises, low lying land near the coast is permanently flooded. The large Teign, Dart and Kingsbridge estuaries in Devon (Fig. 6) have been formed in this way. They are drowned river valleys known as *rias*. In the British Isles the best examples of rias occur in S.W. England and in S.W. Ireland. Notice how the Kingsbridge estuary (Fig. 26) has a branching pattern, the result of the drowning of tributary valleys as well as the valley of the main river. At low tide large areas of mud flats are exposed. In the Dart estuary the mud is 35 m thick in places, showing how much deposition has taken place since the drowning of the river valley.

Rias often occur where the coast is discordant. By contrast, on concordant coasts the valleys which are submerged

Fig 25 A raised beach: Isle of Islay, Scotland

Fig 26 The Kingsbridge Estuary: a ria

Fig 27 Map of the Dalmatian coast

tend to run parallel to the coast. When drowned, such areas are known as *Dalmatian coasts* named after the area of Yugoslavia where a fine example is to be found.

The drowning of glacial valleys produces *fiords* (p. 70). Spectacular examples of these are found in Norway and the sea lochs of N.W. Scotland were formed in the same way.

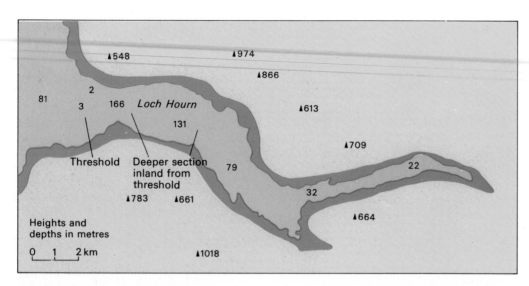

Fig 28 Loch Hourn; a fiord

12 Study the map of Loch Hourn above. Describe the differences between a fiord and a ria, considering both the shape of the inlet as seen from above and the shape of its cross section.

Submergence is also thought to have affected the formation of the *coral islands* which are found in several tropical areas, especially in the Pacific Ocean. Coral consists of the skeletons of *polyps*, small animals which can only live on rocks or coral *reefs* (narrow ridges of coral) within 30 m of the ocean surface. Yet coral is found at depths of over 1000 m, where the polyps could not have survived. Charles Darwin, famous for putting forward the theory of evolution, suggested that the thickness of coral formations could be explained if a coral reef were gradually sinking, with new coral constantly forming at the top. The theory also helps to explain how coral *atolls* – horseshoe or ring shaped islands – may be formed (Fig. 29).

The level of the sea compared with the land is constantly changing and this affects us directly as well as changing the landforms round our coasts. In Finland land has been rising since the melting of the ice cap which once covered it. Some towns which were formerly on the coast are now several kilometres inland and have therefore lost their function as ports. Nearer home, the risk of London being flooded has increased year by year as the land has sunk relative to the sea. The danger of flooding, which would affect over 100 km² and a million people, has led to the construction of the Thames Barrier at Woolwich. Flood gates can rise from the river bed, blocking the Thames and keeping out the dangerous 'storm surges' which occur when high tides coincide with strong northerly winds. It is thought that London will continue to sink and that these gates will have to be used more and more as time passes.

Review Questions

1 Define the terms:
   dominant wave
   discordant coastline
   stack
   tombolo
2 Draw diagrams to show what is meant by:
   longshore drift
   wave refraction
   a ria
3 Explain how:
   a spit is formed
   a coastline may be protected from erosion
   an arch is formed
   you would find out the direction of longshore drift on the beach

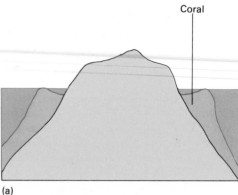

(a)
Island with fringing reef

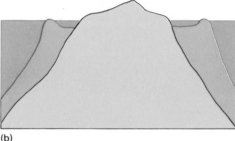

(b)
Island subsiding
Formation of barrier reefs

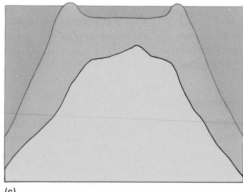

(c)
Subsidence continues
Formation of coral atoll

Fig 29 The formation of coral reefs and atolls

4 Why:
   has the general sea level risen since the Ice Age?
   do tides occur?
   is marram grass sometimes planted on sand dunes?

8 Deserts

Fig 1 A popular image of the desert!

Fig 2 Sandy desert

Fig 3 Rocky desert

The desert environment

1 Before you read this chapter or look closely at the photographs and diagrams, write down ten words that come into your mind when you think of 'deserts'. The most common words could be collected from the class to give a group impression of what deserts are like.

It is likely that words such as 'sand', 'bedouin', 'camels' and 'oases' figured prominently in your list or that of your class. Most people's impressions of deserts are strongly influenced by the sand dunes of the Sahara or by popular images such as in Fig. 1. The Sahara is one of the sandiest deserts in the world and yet as the following figures show, only about a quarter of the desert surface is covered with dunes, most of the desert being rocky or stony.

Type of desert surface	Area of Sahara Desert (%)
desert mountains	45
sand dunes	28
salt crust	12
bare rock floor	10
old volcanoes	3
stony surface	2

The deserts of the world are very dry or *arid*. Some parts of the Atacama Desert in South America for example have never had rainfall recorded. Generally deserts are those areas receiving less than 250 mm of rainfall per year but low rainfall alone does not make a desert. More important is the relationship between rainfall and temperatures as this determines the amount of evaporation that will take place. Deserts are areas where evaporation is greater than rainfall. Many deserts have recorded surface temperatures during the day of up to 70°C although at night the temperatures may fall rapidly.

Rainfall in desert areas is very variable. The average annual rainfall at Biskra, Algeria, for example, is 150 mm yet more than 200 mm fell in a two day storm in September, 1969. Obviously with such freak storms average rainfall figures are very misleading.

Rainfall is also variable from place to place within the desert region. Not all parts of even a small area will get equally wet from the passage of one storm.

2 Study the map of the world's deserts (Fig. 4).

a On a copy of the world map label the following:
   Gobi Desert (Mongolia)
   Sahara Desert (N. Africa)
   Kalahari Desert (S. Africa)
   Arabian Desert
   Iranian Desert
   Thar Desert (India)
   California Desert (N. America)
   Atacama Desert (S. America)
   Gibson and Simpson Deserts (Australia)

b With the exception of the Gobi Desert, which lines of latitude do these deserts lie between?

c Estimate the percentage of the earth's surface that is covered by desert or semi-desert. Your answer should show you why a study of deserts is important to geomorphologists.

d Refer back to Fig. 10 in Chapter 2. In which period of geological time was Britain largely covered by desert?

Weathering and the desert surface

Deserts are very varied in appearance. The great seas of sand in N. Africa and Arabia are known as *erg* desert (see Fig. 2). The name *reg* is often given to the vast stony surfaces of N. Africa where wind has blown away finer sand leaving a barren expanse of pebbles (Fig. 3). The evaporation of ground water produces hardened crusts of salt near the surface. Wind and weathering combine to bring a range of reddish colours and strange shapes to the desert landscapes but many of the landforms, although striking in appearance are quite rare. They have often become well-known simply because they stand out in an otherwise featureless environment.

When rocks are heated by the sun they expand. As they cool they shrink. The strain causes rocks to split. This is a type of physical weathering (see Fig. 3). Other types of weathering do take place in deserts however, although as there is less water present, chemical weathering is not as obvious as in more humid areas. Nevertheless, the small quantity of water that there is plays an essential part in the weathering process in deserts. Dew often forms near the ground and so chemical weathering may be active in hollows or at

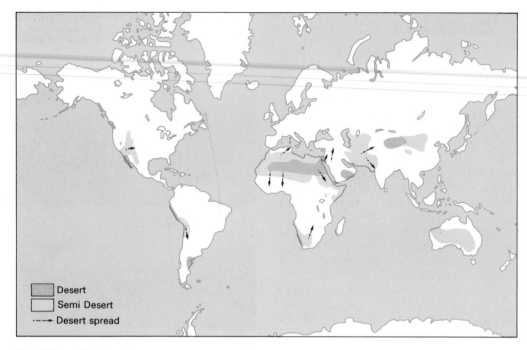

Fig 4 The world's deserts

Desert

Semi Desert

- - -> Desert spread

the base of rock outcrops. This can cause the weathering of softer bands of rock near the ground sometimes forming *mushroom rocks*.

Another type of weathering important in hot deserts is the growth of salt crystals in rock. As the desert temperatures are high, water in the rocks evaporates rapidly leaving crystals of salt to grow near the surface.

As the crystals expand they force the rocks apart in the same way ice does in freeze/thaw weathering.

Generally, as weathering is slower in deserts, the layer of weathered rock fragments (the *regolith*) tends to be shallower.

Fig 5 Mushroom rocks

Fig 6 A wadi: Tunisia

Water in deserts

Although there is little water in deserts many of the features of erosion are formed by water. Many deserts have had wetter climates in the past and today's landforms were shaped when there was more water available.

Rivers in deserts differ from those in more humid environments in several respects. Firstly, the nature of desert rainfall means that the flow of rivers is often irregular. Streams flow when there is rainfall and dry up when there is not. These streams are known as *intermittent streams*. In the event of the freak storms described earlier, torrents of water may rush down valleys which are otherwise dry for many months. Dry valleys that experience these infrequent flows of water are termed *wadis*.

As evaporation rates are high many rivers do not reach the sea but evaporate in great inland depressions or *playas*. These depressions may cover large areas and become lakes of mud after a sudden storm. They receive not only the solid load of rivers but the dissolved chemicals as well. When the water evaporates, the mud dries to produce sun cracks, often in the shape of polygons. Layers of salt are also deposited.

Although desert scenery is very varied many deserts share the features (shown in Fig. 7), of a mountain mass, gently sloping rock surface (or *pediment*) and *alluvial fans*.

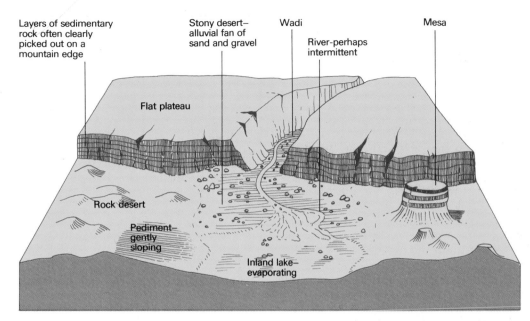

Fig 7 Some desert landforms

Labels: Layers of sedimentary rock often clearly picked out on a mountain edge; Stony desert—alluvial fan of sand and gravel; Wadi; River—perhaps intermittent; Mesa; Flat plateau; Rock desert; Pediment—gently sloping; Inland lake—evaporating

3 Study Fig. 7 and consider the following facts;
When a stream passes from an upland area to a lowland area its gradient is reduced
River gradient is related to the speed of river flow
The speed of a river determines the amount of load it can carry
Describe in detail how alluvial fans form.

Pediments appear to be left as the edge of the mountain mass is eroded back. As this happens, remnants of the mountains or plateaux are often left standing.

4 Study the photograph, Fig. 9 and the enlargement on p. 1. This shows all that remains of a plateau surface in Monument Valley, Utah, U.S.A. The flat-topped remnants of the plateau are called *mesas* and those that are too small to have preserved a flat top are called *buttes*.

Draw a mesa and a butte, labelling:
i) the sedimentary rock strata ii) the screes iii) the differences between the two landforms.

Fig. 8 shows Ayers Rock in central Australia. Once a sacred place for aborigines, this sandstone mountain, 350 m high, is now a popular tourist attraction. It is an *inselberg* – an upstanding mass of resistant rock rising sharply above the surrounding *pediment*. It is likely that inselbergs are formed in the same way as mesas.

Fig 8 Ayers Rock, Australia

Fig 9 Monument Valley, Utah, USA

Landforms produced by wind

5 Fill a shallow tray with fine, dry sand. Blow across the surface of the tray. Moisten the sand and blow across the tray again. Describe what happens in both cases.

The wind acts as an agent of erosion, transportation and deposition in deserts. You may have seen the outside of buildings in cities being cleaned of their surface coating of grime by *sandblasting*. This method of cleaning uses jets of air with grains of sand in them. The air and sand is directed onto the stonework and the sand scratches away the dirt. Desert winds act as agents of erosion in much the same way. Travellers who have left cars in desert sandstorms have returned to find the paint stripped off by the wind exposing the bare metal beneath. Stones on the desert surface may also be polished by the sandblasting. The distinctive pebbles so formed are called *ventifacts*.

It is likely that the action of sandblasting (together with chemical weathering near the surface) helps the formation of mushroom rocks. How much of the shaping of these features is due to chemical weathering and how much to wind erosion probably varies from one place to another. The wind however seems to be more important in the formation of *yardangs*, (see Fig. 10), as these elongated ridges of rock lie parallel to the direction of the wind.

The Qattara depression in the Sahara is a saucer-shaped depression about 300 by 150 km and about 125 m below sea level in the centre. Depressions such as these collect water and therefore rock breakdown by chemical weathering is more common than elsewhere in the desert. The wind blows away the dry, loosened sand and rock waste and the hollow is deepened. This process is called *deflation*. The Qattara depression is a large *deflation hollow*.

6 Study Fig. 11.

a Compare the movement of sand particles by the wind with the transport of load by a river. Mention the size of the particles, the methods and the names give to the processes of transportation.

b Describe the flight path of sand grains in the zone of saltation.
c Look at the illustrations of yardangs and mushroom rocks. At what height above ground will erosion be greatest?

The speed of the wind determines the amount of sand it can transport. If small patches of sparse vegetation slow down the wind speed, the wind cannot hold all the sand and so some is deposited. Sand dunes often begin like this.

Dunes are not a haphazard piling up of sand. They often have regular patterns of size, spacing and shape. Sand shapes are usually of three main sizes. *Ripples* are common everywhere in sandy deserts and are a few centimetres to a metre in height. *Dunes* are from about a metre to about 30 m high. They tend to be about 100 to 200 m apart. The largest features are *draa*, up to 300 m high and a kilometre or two apart.

Sand dunes commonly form a regular pattern known in the Sahara as *aklé*. This pattern is shown in Fig. 12. Wind eddies scoop out hollows and spread out tongues of sand. Less common, but more spectacular are *seif dunes* and *barchans*.

Fig 10 Yardangs

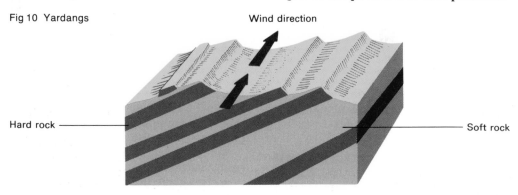

Fig 12 Aklé pattern of sand dunes

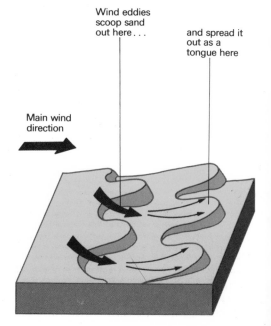

Fig 11 The transport of sand by wind

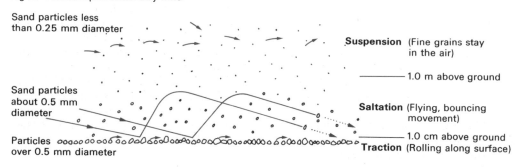

90

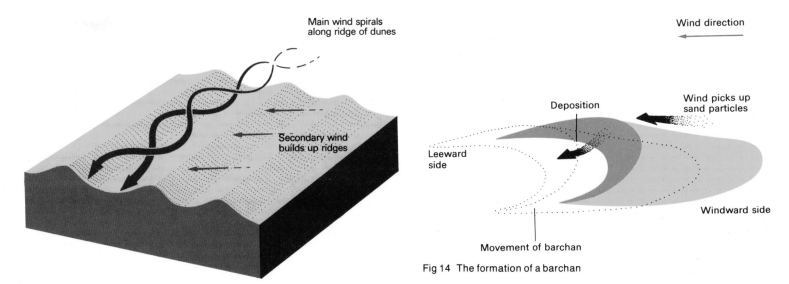

Main wind spirals
along ridge of dunes

Secondary wind
builds up ridges

Fig 13 The formation of seif dunes

Wind direction

Deposition

Wind picks up
sand particles

Leeward
side

Windward side

Movement of barchan

Fig 14 The formation of a barchan

Seif dunes form in desert areas where the wind blows mainly from one direction but where there are cross winds. The barchan dune is quite rare and tends to form in areas where the wind blows from one direction only. Fig. 14 shows how this type of dune moves.

People and the desert

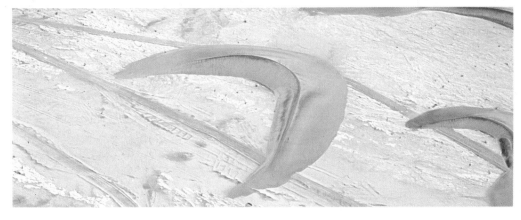

Fig 15 A barchan: the Namib Desert

Consider this extract from an arabic dictionary:

Thirst :	al'atash	thirst
	al-Zama'	thirst
	al-Sada	severe thirst
	al-Ghulla	burning thirst
	al-Luhba	burning thirst
	al-Huyam	violent thirst
	al-Uwam	burning thirst, causing giddiness
	al-Juwad	terrible thirst, causing death.

In very dry countries, the attitude of the people towards water is very different from ours. People who are frequently thirsty can recognise degrees of thirst that we are unaware of. As the population increases in arid areas greater concern is felt for water

supply. Israel for example passed a law in 1948 giving the state full ownership of the water resources. In other words, water was nationalised.

Most settlement in deserts occurs along river banks or at oases. Rivers flowing through deserts often have a marked seasonal pattern of flow causing flooding at particular times of the year and low water at other times.

7a Using an atlas draw a sketch map to show the following features:
the River Nile and its headwaters (the Blue Nile, White Nile and Atbara rivers).
The Red Sea and the Egyptian Coast
The area of reeds and marsh in the Sudan known as the Sudd.

Contours at approximately 200, 400, 1000, 1500 and 3000 m.

b Using your map and the graph of the annual flow of the River Nile (Fig. 16), answer the following:
Where is the source of the White Nile?
Which river has the steeper gradient – the White Nile or the Blue Nile?
When does the Nile flood?
Which river contributes most to the flow of the Nile in spring?
Which river contributes most to the flow of the Nile in autumn?

c Dams have been built on the River Nile to prevent flooding and to provide a store of water for release during the dry months. From your atlas locate these dams and add them to your map.

Settlements also occur at *oases*.

8 Revise the meanings of the following words:
syncline
aquifer
artesian basin

Many deserts are in areas of sedimentary rock in the form of large synclines. If these strata are composed of alternate permeable and impermeable layers a situation similar to that in the London Basin will occur. Oases are places in the desert where water can be obtained. They are often found where deflation has lowered the surface of the desert to the point where aquifers may be tapped such as at the Siwa oasis in the Qattara depression.

Fig 17 Desertification is a vicious circle

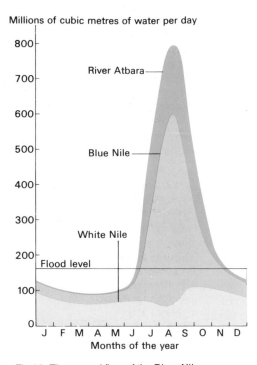

Millions of cubic metres of water per day

Fig 16 The annual flow of the River Nile

Desertification

The word desertification has been given to the process of desert spread.

9 Look at Fig. 4 on p. 88.
List the areas in the world where the deserts are spreading.

Desertification is causing increasing anxiety in the countries where it occurs. The photograph from the Middle East above shows how desertification might be caused. A change in the climate can bring about lower rainfall but at many desert edges it is man himself who has caused the deserts to spread. Often the population has increased. More food is needed and so farmers cannot afford to leave land fallow to replenish the nutrients that have been lost to the crops. This results in the soil becoming rapidly exhausted. It is less able to support further crops and without a plant cover the soil cannot hold water so easily. As we have seen, a dry soil is easily eroded by the wind. This happened in the Dust Bowl of the United States in the 1920s and 1930s and is happening today in the 'Sahel' – the Southern edge of the Sahara desert.

With a growing population, herdsmen have less space to graze their flocks of sheep and goats. They must return to areas that have been grazed without allowing enough time for the vegetation to grow. Overgrazing and the trampling of the animal's hooves also causes the soil to deteriorate, leaving it more susceptible to wind erosion. The spread of desert sand and the loss of vegetation means that rainfall runoff is more rapid, less water is trapped in the soil and makes future vegetation growth unlikely. Such a chain of events, gradually becoming worse, is called a *vicious circle*.

There are no simple solutions to the problems of desertification. Simply supplying water to the dry areas may present further problems. Much of the water in deserts is slightly salty, and when used for years to irrigate crops may eventually, through its evaporation, cause a build up of salts which renders the soil useless for agriculture.

Desertification is an appropriate subject with which to end this book. It illustrates that people live in a very close relationship with their environment and that this relationship is often a delicate balance that may easily be upset. Geomorphologists, in attempting to understand the processes at work in deserts are, together with other scientists, working towards solutions.

Index

Acknowledgements

The publishers would like to thank the following for permission to reproduce photographs:

Aerofilms, pp.2, 24, 29, 40, 44, 68, 69, 79, 85, 89; Ardea, pp.35, 89; K.S. Blake, pp.29, 36; Bridgeman Art Library, p.64; Bruce Coleman, pp.88, 89, 91; Department of Geology, Oxford University, p.7; Geoscience Features, pp.27, 87; Geoslides, p.87; Susan Griggs Agency, p.92; Institute of Geological Sciences, pp.7, 8, 19, 23, 25, 28, 29, 30, 31, 63; Eric Kay, pp.5, 24, 29, 31, 32, 45, 54, 66, 72, 73, 80; Keystone Press Agency, p.42; National Museum of Antiquities of Scotland, p.16; Norwegian National Tourist Office, p.70; Oxford University Cave Club/Simon Fowler, pp.39, 40; Solarfilma, p.63; Sporting Pictures, p.23; Surrey Comet, p.50; United States Geological Survey, p.30; University of Leeds, pp.9, 12; Vision International/Krafft Explorer p.33.

Cover illustrations and figures 1.4, 2.1, 2.10, 3.2, 3.20, 3.37, 4.2, 4.12, 4.23, 4.24, 5.5, 5.22, 6.6, 6.11, 6.13, 6.22 and 6.23 by Gary Hincks.